STUDY GUIDE for SOCIOLOGY: A GLOBAL PERSPECTIVE

Second Edition

Joan Ferrante
Northern Kentucky University

Wadsworth Publishing Company
I(T)P™ An International Thomson Publishing Company

Belmont • Albany • Bonn • Boston • Cincinnati • Detroit • London • Madrid • Melbourne
Mexico City • New York • Paris • San Francisco • Singapore • Tokyo • Toronto • Washington

Printed in the United States of America
1 2 3 4 5 6 7 8 9 10—01 00 99 98 97 96 95

For more information, contact Wadsworth Publishing Company.

Wadsworth Publishing Company
10 Davis Drive
Belmont, California 94002 USA

International Thomson Publishing Europe
Berkshire House 168-173
High Holborn
London, WC1V 7AA England

Thomas Nelson Australia
102 Dodds Street
South Melbourne 3205
Victoria, Australia

Nelson Canada
1120 Birchmount Road
Scarborough, Ontario
Canada M1K 5G4

International Thomson Editores
Campos Eliseos 385, Piso 7
Col. Polanco
11560 México D.F. México

International Thomson Publishing GmbH
Königswinterer Strasse 418
53227 Bonn Germany

International Thomson Publishing Asia
221 Henderson Road
#05-10 Henderson Building
Singapore 0315

International Thomson Publishing Japan
Hirakawacho Kyowa Building, 3F
2-2-1 Hirakawacho
Chiyoda-ku, Tokyo 102 Japan

ISBN 0-534-20977-7

Contents

Preface P.6

1 The Sociological Imagination 1.1
- Study Questions 1.1
- Concept Application 1.4
- Applied Research 1.6
- Practice Multiple-Choice and True/False Questions 1.6
- Continuing Education 1.9
  - Leisure 1.9
  - Travel 1.10
  - General and Readable Information 1.10
  - Chapter References 1.11
  - Answers 1.11

2 Theoretical Perspectives 2.1
- Study Questions 2.1
- Concept Application 2.6
- Applied Research 2.9
- Practice Multiple-Choice and True/False Questions 2.9
- Continuing Education 2.16
  - Leisure 2.16
  - Travel 2.16
  - General and Readable Information 2.17
- Country Profile: Mexico 2.17
- Chapter References 2.19
- Answers 2.19

3 Research Methods 3.1
- Study Questions 3.1
- Concept Application 3.8
- Applied Research 3.10
- Practice Multiple-Choice and True/False Questions 3.11
- Continuing Education 3.19
  - Leisure 3.19
  - Travel 3.20

General and Readable Information 3.20
Country Profile: Japan 3.21
Chapter References 3.22
Answers 3.23

4 Culture 4.1
Study Questions 4.1
Concept Application 4.5
Applied Research 4.8
Practice Multiple-Choice and True/False Questions 4.8
Continuing Education 4.15
Leisure 4.15
Travel 4.15
General and Readable Information 4.16
Country Profile: South Korea 4.16
Chapter References 4.17
Answers 4.18

5 Socialization 5.1
Study Questions 5.1
Concept Application 5.5
Applied Research 5.7
Practice Multiple-Choice and True/False Questions 5.8
Continuing Education 5.15
Leisure 5.15
Travel 5.15
General and Readable Information 5.16
Country Profile: Israel, West Bank, and Gaza 5.16
Chapter References 5.18
Answers 5.19

6 Social Interaction and the Social Contruction of Reality 6.1
Study Questions 6.1
Concept Application 6.6
Applied Research 6.10
Practice Multiple-Choice and True/False Questions 6.10
Continuing Education 6.17
Leisure 6.17

Travel 6.17
General and Readable Information 6.18
Country Profile: Zaire 6.18
Chapter References 6.20
Answers 6.20

7 Social Organizations 7.1
Study Questions 7.1
Concept Application 7.5
Applied Research 7.8
Practice Multiple-Choice and True/False Questions 7.8
Continuing Education 7.14
Leisure 7.14
Travel 7.15
General and Readable Information 7.15
Country Profile: India 7.15
Chapter References 7.16
Answers 7.17

8 Deviance, Conformity, and Social Control 8.1
Study Questions 8.1
Concept Application 8.7
Applied Research 8.9
Practice Multiple-Choice and True/False Questions 8.10
Continuing Education 8.16
Leisure 8.16
Travel 8.16
General and Readable Information 8.17
Country Profile:China 8.17
Chapter References 8.18
Answers 8.19

9 Social Stratification 9.1
Study Questions 9.1
Concept Application 9.4
Applied Research 9.7
Practice Multiple-Choice and True/False Questions 9.8
Continuing Education 9.14

Leisure 9.14
Travel 9.14
General and Readable Information 9.15
Country Profile: South Africa 9.15
Chapter References 9.18
Answers 9.18

10 Race and Ethnicity 10.1
Study Questions 10.1
Concept Application 10.4
Applied Research 10.7
Practice Multiple-Choice and True/False Questions 10.7
Continuing Education 10.14
Leisure 10.14
Travel 10.14
General and Readable Information 10.14
Country Profile: Germany 10.15
Chapter References 10.16
Answers 10.16

11 Gender 11.1
Study Questions 11.1
Concept Application 11.4
Applied Research 11.7
Practice Multiple-Choice and True/False Questions 11.7
Continuing Education 11.13
Leisure 11.13
Travel 11.14
General and Readable Information 11.14
Country Profiles: Bosnia and Herzgovina, Croatia, Serbia and Montengro, Slovenia 11.15
Chapter References 11.19
Answers 11.19

12 Population and Family Life 12.1
Study Questions 12.1
Concept Application 12.6
Applied Research 12.8

Practice Multiple-Choice and True/False Questions 12.9
Continuing Education 12.14
Leisure 12.14
Travel 12.15
General and Readable Information 12.15
Country Profile:Brazil 12.15
Chapter References 12.17
Answers 12.18

13 Education 13.1
Study Questions 13.1
Concept Application 13.5
Applied Research 13.8
Practice Multiple-Choice and True/False Questions 13.8
Continuing Education 13.15
Leisure 13.15
Travel 13.16
General and Readable Information 13.17
Country Profile: United States of America 13.17
Chapter References 13.19
Answers 13.19

14 Religion 14.1
Study Questions 14.1
Concept Application 14.5
Applied Research 14.8
Practice Multiple-Choice and True/False Questions 14.8
Continuing Education 14.13
Leisure 14.13
Travel 14.14
General and Readable Information 14.15
Country Profile: Lebanon 14.15
Chapter References 14.16
Answers 14.17

15 Social Change 15.1
Study Questions 15.1
Concept Application 15.6

Applied Research 15.9
Practice Multiple-Choice and True/False Questions 15.10
Continuing Education 15.16
Leisure 15.16
Travel 15.17
General and Readable Information 15.17
Chapter References 15.18
Answers 15.18

Appendix A A.1
How to Find Sources for Research Papers A.1
How to Write a Research Paper A.6
How to Write a Book Review A.13

Appendix B B.1
Maps

Preface

I wrote the *Study Guide* for *Sociology: A Global Perspective* hoping to make it more than simply a test preparation tool. Everyone intuitively knows that memorizing material for the sake of doing well on a test cannot lead to meaningful and long-lasting learning experiences. In other words, if students read the text with an eye toward predicting possible test questions and take notice on lectures without thinking about them until the night before a test, the academic experience will be an empty one. Real learning means thinking about the ideas that you hear and read, taking an active role in learning, and incorporating learning experiences into the activities of daily life. I wrote the Study Guide with this vision of learning in mind. Each chapter corresponds to a chapter in the textbook and includes six selections: (1) study questions, (2) concept application, (3) applied research, (4) practice multiple-choice and true/false questions, (5) continuing education, and (6) country profile. There is also an appendix A to the guide prepared by Paul Knepper, Justice Studies Program, Northern Kentucky University and an Appendix B composed of maps.

Study Questions

Think of the study questions as note-taking tools. If you answer these questions thoughtfully and conscientiously, you will come away with a thorough synthesis of the sociological material covered in the textbook.

Concept Applications

Each chapter includes five concept-application scenarios. You are asked to determine which concept or concepts is best represented in that scenario and to explain why.

Applied Research

These are mini-research projects that promote active learning because they require you to find answers to questions that are related to topics discussed in each chapter. Mostly they are projects that can be completed in an afternoon.

Practice Multiple-Choice and True/False Questions

If you treat these questions as a comprehensive study tool, you will not be adequately prepared to take a test. On the other hand, if you prepare beforehand (as if you are going to take a test that will count for a grade), the score you earn will probably be a good indicator of how well you can expect to do on the actual test.

Continuing Education

This selection offers ideas that can help make learning a part of life now and after college. There is a leisure activity section (which includes a short list of music and movie suggestions), a travel section (which suggests the names of various exhibits, museums, galleries, and other centers of historic, scientific, cross-cultural, and social value to visit in the United States), and a recommended list of periodicals, magazines, and newspapers about or from the countries emphasized in the textbook.

Country Profiles

The country profiles are excerpted from the 1994 *World Population Data Sheet* published by the Population Reference Bureau; *Background Notes* published by the U.S. Department of State; and from *The World Fact Book 1993* published by the U.S. Central Intelligence Agency. The reports give additional background on various countries emphasized in *Sociology: A Global Perspective*.

Appendix A and B

Appendix A consists of three articles designed to make the learning process a more productive and efficient experience: "How to Find Sources for Research Papers," "How to Write a Research Paper," and "How to Write a Book Review." Appendix B contains maps of each country emphasized in the textbook.

Chapter 1
The Sociological Imagination

Study Questions

1. What is the sociological imagination?

2. Distinguish between troubles and issues. Why do people often feel trapped by issues?

3. Use the case of global interdependence to illustrate some of the ways issues are experienced on a personal level.

4. According to C. Wright Mills, why is merely keeping up with information about events in the world not enough to help people see what is happening within themselves?

5. Explain the following statement: "If only by living in a society, people shape that society, however minutely, even as the society shapes them." (Refer to the case of Phil Voss.)

6. Why is it often difficult to determine what is an "American-made" product?

7. What is social relativity? Give an example.

8. What is the transformative power of history? As examples, use the Industrial Revolution and its effects on the nature of work and interaction.

9. Define Karl Marx's, Emile Durkheim's, and Max Weber's vision of what the task (or central concern) of sociology should be.

10. Who is Harriet Martineau? How did she conduct research for *Society in America*? What methods did she use to make sense of her observations?

11. What does Josh Gibson's case teach us about the personal and the larger consequences of ignoring some people's contributions or talents?

12. What larger historical forces shaped the foundation of sociology? What kinds of events are shaping the need to study sociology from a global perspective?

13. Why are the data on interconnection placed at the beginning of each chapter?

14. In the broadest sense of the word, what does the adjective *American* apply to? How has that adjective come to be associated with the United States?

Concept Application

Below are five scenarios and the sources from which they were drawn. Decide which concept or concepts covered in Chapter 1 are represented best by each scenario, and explain why. The following concepts are considered:

Bourgeoisie
Disenchantment
Global interdependence
Issues
Means of production
Mechanization
Proletariat
Social action
Social relativity
Sociological imagination
Solidarity
Transformative power of history
Troubles

Scenario 1

"Flip channels around these days and it's possible to encounter broadcast programming in Spanish, Arabic, Cantonese, Hindi, Tagalog, Cambodian, Vietnamese, German, Pusthu, Hungarian, Thai, French, Hebrew, Italian and more. The TV set has become a place to catch up with news from your native country, to listen to programming in your native tongue or to simply try to understand another language as it's spoken" (Mendoza 1993, p.6).

Scenario 2

"The United States Postal Service plans to test a kind of do-it-yourself post office....The machines, to be placed in post office lobbies, will combine an advanced scale to weigh letters and packages and calculate postage, with a machine that will print stamps in the exact denomination needed after deposit of the proper amount (it even gives change)" (*The New York Times* 1992, p. 34Y).

Scenario 3

" 'Everyone in Reno fell in love with the Russians because they were so warm, unassuming, and generous,' recalls Dorothy Souza, the faculty visitors coordinator in charge of international programs.

She continues, 'When the Russians arrived at Reno Airport, they had an incredible amount of luggage. We thought their bags were packed with clothes. Instead, they were filled with modest gifts, which they gave to everyone they met.'

The American hosts discovered the Russians preferred to be call 'Mr.' instead of 'Judge' and that, yes, there is pizza in Russia.

'One night we took the Russians out for pizza and ice-cold glasses of beer,' says Souza. 'But they sent the beer back--to be heated' " (Roessing 1992, p. 26).

Scenario 4

"The wheels of our chartered Aeroflot TU154 bounced onto the runway. I looked out the window for my first glimpse of Petropavlovsk-Kamchatskiy, the supersecret city and submarine port on Siberia's Kamchatka Peninsula, but saw only swirling snow and darkness....

Even a very few years ago, this trip would have been impossible; the border between mainland Siberia and Alaska, only 57 miles apart at the Bering Strait, had been frozen by the Cold War since 1948. Not until the *glasnost* of the late 1980s did the two neighbors begin to think they could be more than the mysteries on the horizon. Tentative messages began to flow in 1987; the thaw officially started in June 1988 when a planeload of officials and Alaska natives flew from Nome to Provideniya, across the strait, for a one-day visit.

Since then the subarctic flirtation has blossomed into a serious romance, with small but steadily increasing numbers of businessmen, bureaucrats, scientists and ordinary people making their way across the icy waters, despite difficulties of communication and transportation and the Soviets' lack of hard currency" (Shute 1991, p. 31).

Scenario 5

"The world's trade in bananas is dominated by just three huge food multinationals: United Brands (with a 34 percent market share in 1974), Standard Fruit (with a 23 percent) and Del Monte (10 percent). As with many other commodities, the companies control the transport, packaging, shipment, storage and marketing of the fruit. As a result, the profits from bananas go largely into western pockets, while the producer countries get only a pittance" (Harrison 1987, p. 348).

Applied Research

Find a newspaper article or book review in which the author connects an incident in the life of an ordinary individual to a larger historical force. Focus on how that person's response shapes his or her society, however minutely, even as that society shapes him or her.

Practice Test

Multiple-Choice Questions

1. People possess the sociological imagination when they can do which one of the following?
   a. Connect seemingly impersonal and remote historical forces to the most basic incidents of an individual's life.
   b. Identify the personal characteristics or immediate relationships that cause issues.

c. Act as spectators to the larger forces that affect their lives.
d. Define an issue as it relates to shortcomings in individual character.

2. Unemployment is ______ when it is caused by fundamental large-scale changes in the economy, such as plant relocations.
   a. trouble
   b. an issue
   c. relative
   d. uncontrollable

3. During the Middle Ages, the building of a cathedral took
   a. one or two decades.
   b. a generation.
   c. several lifetimes.
   d. about 10 years.

4. At some level, the world has been interdependent
   a. for at least 500 years.
   b. since 1850.
   c. since 1960.
   d. since the early 1980s.

5. One fundamental feature of the Industrial Revolution is
   a. craftsmanship.
   b. solidarity.
   c. manual labor.
   d. mechanization.

6. The Industrial Revolution transformed the nature of work in which one of the following ways?
   a. Machine production was replaced by hand production.
   b. People now could say, "I made this; this is a unique product of my labor."
   c. Products became standardized, and workers performed specific tasks in the production process.
   d. Artisans' power over the production process increased dramatically.

7. In the most general sense, sociology is
   a. the disciplined and objective study of social interaction.
   b. the study of the industrialization process.
   c. the study of the causes and consequences of inequality.
   d. the disciplined study of the course and consequences of industrialization

Use one from the following set of responses for questions 8-15 to identify the thinker associated with each statement.

a. Karl Marx
b. Emile Durkheim
c. Max Weber
d. Harriet Martineau

8. The sociologist's task is to analyze and explain the ties that bind people to one another.

9. Every historical period is characterized by a system of production that gives rise to specific types of confrontation between an exploiting and an exploited class.

10. In conducting social research it is important to see a country in all its diversity.

11. In industrial societies, people are united by the fact that they cannot earn their livelihoods independent of one another.

12. The sociologist's task is to analyze and explain the reasons for and the course and consequences of social action.

13. Every aspect of people's lives--their occupation, their religious beliefs, their level of education, how they spend their leisure time--personifies the social class to which they belong.

14. Compare actual workings of society with the principles on which the society was founded.

15. With industrialization, behavior is less likely to be guided by tradition or emotion and more likely to be value-rational.

16. Which of the following statements is true about the role of the United States in world affairs?
    a. The United States is the only economic power in the world.
    b. The United States, the People's Republic of China, and Russia are the only countries with nuclear weapons.
    c. The capacity of the United States to control international affairs has declined since the end of World War II.
    d. By all measures, the United States is in a state of decline.

True/False Questions

T F 1. Social relativity and the transformative powers of history are unrelated concepts.

T F 2. The Industrial Revolution affected almost every aspect of daily life.

T F 3. Few insights about the character of contemporary American society can be gained from reading the writings of Marx, Weber, and Durkheim, who wrote before 1920.

T F 4. For the most part, education in the United States is structured in such a way that understanding life in other countries, especially non-Western countries, is an important element of the curriculum.

T F 5. The United States is the only industrialized country in the world in which students are not required to learn a second language.

Continuing Education

Leisure

For a change in music, listen to

Voices: A Compilation of the World's Greatest Choirs. Mesa.
Collections of National Anthems: Europe/USSR/Asia Vol. 1. Denon.
Collections of National Anthems: Americas/Africa/Middle East. Oceania.

The next time you rent a movie, consider

Roger and Me, directed by Michael Moore (91 minutes). The film documents how the people living in Flint, Michigan, were affected by General Motors executives' decisions during the 1980s to close 11 North American factories and to lay off 30,000 autoworkers (with a specific focus on Roger Smith, chairman of GM). The film is especially timely in light of the December 1991 announcement that GM planned to eliminate several auto plants and 74,000 jobs by 1995. As you view the film, consider the distinction between troubles and issues. List the ways in which various individuals in the film respond. What happens to the city of Flint? How do the well-off respond to the plight of the unemployed? Do they see the problems in Flint as troubles or as issues?

Travel

If you're in Seattle, visit the Museum of Flight (9404 E. Marginal Way S.; 206-764-5720). The museum is housed next to the Boeing Corporation's test field and contains approximately 30 aircraft dating from 1916 to the present. It also has a gallery documenting manned outer space flights.

If you're in St. Louis, visit the National Museum of Transport (3015 Barrett Station Road; 314-965-7988). This museum contains every kind of ground transport that has been invented over the past 200 years.

If you're in Sante Fe, visit the Museum of International Folk Art (706 Camino Lejo; 505-827-8350) which houses one of the largest collections of folk art from countries around the world.

General and Readable Information

Population Today (1875 Connecticut Avenue NW, Suite 520, Washington, DC 20009-5728) is a monthly newsletter published by the Population Reference Bureau. It includes short articles on population issues faced by countries around the world. Recent articles include "Infant Mortality: Who's Number One?" and "World's Largest Head Count Ever: A Look at the Numbers for the 1990 China Census."

The UNESCO Courier (31, rue Francois Bonvin, 75015 Paris, France) is an international cultural magazine published monthly in 35 languages and braille, and read in 120 countries. The magazine is published monthly, and it explores topics as they relate to countries around the world.

The World Paper (210 World Trade Center, Boston, MA 02210) is a monthly international publication of news and views. The writers are natives of the countries about which they write. *The World Paper* appears as a special section in newspapers and magazines in 22 countries, but also can be purchased through subscription.

Chapter References

Harrison, Paul. 1988. *Inside the Third World: The Anatomy of Poverty*, 2nd ed. New York: Viking Penguin.

Mendoza, N.F.1993. "International House of TV." *Los Angeles TV Times* (August 8):6-7.

The New York Times. 1992. "Newest Postal Clerks Won't Smile or Frown." (June 20): 34Y.

Roessing, Walter. 1992. "Order, Order." *Sky* (March):22-26.

Shute, Nancy. 1991. "From Unalaska to Petropavlovsk: Warm Welcomes Amid Geysers and Snow." *Smithsonian* (August):30-39.

Answers

Concept Application

1. Global interdependence
2. Mechanization
3. Social relativity
4. Transformative power of history
5. Means of production; Bourgeoisie

Multiple-Choice

1.a 2.b 3.c 4.a 5.d 6.c 7.a 8.b 9.a 10.d 11.b 12.c 13.a 14.d 15.c 16.c

True/False

1.F 2.T 3.F 4.F 5.T

Chapter 2
Theoretical Perspectives

Study Questions

1. Under what circumstances are facts valuable and useful?

2. What is the popular (very common) interpretation of the facts listed at the beginning of Chapter 2?

3. What is the major goal of the NAFTA agreement that went into effect January 1, 1994?

4. How should we think about NAFTA's potential effects on society?

5. What is the *maquiladora* program? Why was it established?

6. What central questions would a functionalist, a conflict theorist, and a symbolic interactionist ask about the *maquiladora* program?

7. What is a function? As examples, consider the functions of (a) education; (b) unwritten rules about appropriate times to communicate.

8. What are the major shortcomings of the functionalist perspective?

9. According to the functionalist perspective, why has poverty not been eliminated?

10. What concepts did Robert K. Merton introduce to counter criticisms of the functionalist perspective? Briefly define each concept and explain how they are combined to refine the perspective. What criticism is not addressed by Merton's concepts?

11. Use the following chart to summarize a functionalist analysis of the 1986 Immigration Reform and Control Act.

	Function	Dysfunction
Manifest		
Latent		

12. Consider the manifest and latent functions and dysfunctions associated with the *maquiladora* program. List those which you believe also might apply to the NAFTA agreement.

13. Explain the following statement: "The manifestations of a common border culture are found less in formal agreements than in activities of thousands of people who divide their lives between the two sides."

14. Is women's employment in the *maquiladora* program a latent function or a dysfunction? Explain.

15. Distinguish between the bourgeoisie and the proletariat. Which of the two is the exploiting class? Why? How is the exploitation justified?

16. Identify the facade of legitimacy that some people use to explain the fate of jobless workers. How would a conflict theorist react to such justifications?

17. What are the major shortcomings of the conflict perspective?

18 State how each concept listed below applies to the *maquiladora* program:

Means of production

The bourgeoisie

The proletariat

The facade of legitamcy versus "reality"

19. What is a symbol?

20. Explain the following statement: "Meaning or value attached to a physical phenomenon is not evident from the physical phenomenon alone."

21. What are the major shortcomings of the symbolic interactionist perspective?

22. Which symbols did American marketers and leaders from border communities use to counter negative images of Mexico?

Concept Application

Below are five scenarios and the sources from which they were drawn. Decide which concept or concepts covered in Chapter 2 is represented best by each scenario, and explain why. The following concepts are considered:

Dysfunctions
Facade of legitimacy
Function
Ideologies
Latent dysfunctions
Latent functions
Manifest dysfunctions
Manifest functions
Sociological theory
Symbol
Theory

Scenario 1

"The influx of Korean-owned firms conferred obvious economic benefits on Los Angeles. (1) Korean firms tended to service low income, nonwhite neighborhoods generally ignored and underserved by big corporations....(2) The Korean influx restored the [deteriorating and underutilized] neighborhoods in which Koreans settled....(3) Their residential and commercial interests compelled Koreans to combat street crime....(4) Koreans valued public education and improved it. Indeed, many Korean families had emigrated to the United States because of this country's superior educational opportunities" (Light and Bonacich 1988, pp. 6-7).

Scenario 2

"For the Korean worker in the 1960s the situation was grim. $80-$225 per capita national income equals misery for most everyone. Workers lived in overcrowded sections of town near the factory in which they worked....[A] small minority was gaining exorbitant wealth while the [majority of workers was] asked to sacrifice for the good of the nation. A story illustrates the point.

It is often the practice of company presidents to visit one of the their factories, gather all the employees together and exhort them to make greater efforts. In one instance the company president had just returned from the United States where he had completed a deal to export some of the company's products to America. He spoke to the workers in this vein: 'I was able to secure this contract with the American firm only because our products are cheap. And our products are cheap only because of your hard work and low wages. You are the true builders of our nation, the true patriots'" (Ogle 1990, pp. 19-21).

Scenario 3

"Stereotypes about Tijuana and other Mexican border cities endure on both sides. They are seen as tawdry, artificial outposts whose Mexican identity has been diluted by a longtime bombardment of U.S. culture: television, McDonald's and rock music. And economic integration accelerated by the North American Free Trade Agreement has intensified fears that these ills are spreading south.

Tijuana has been further battered by the assassination of presidential candidate Luis Donaldo Colosio last month. The ensuing worldwide headlines depicted once more the traditional sinister images of lawlessness and violence" (Rotella 1994, p. A1).

Scenario 4

"Dr. Louise Keating became 'Trash Czar' for a few days. Dr. Keating, director of Red Cross Blood Services in Cleveland, found her center almost engulfed by mounds of debris--dressings, needles, plastic tubes--most of it the usual detritus of any organization, but some of it splashed with the blood of donors. Her center was not generating any more trash than usual. But suddenly no one was willing to cart it away. AIDS could be transmitted through blood, we had now learned. Last year's innocuous garbage had become this year's plague vector. Or so it seemed to Cleveland's carters. And the refuse piles grew.

Dr. Keating did solve her problem. Now, all waste that has any blood on it is sterilized in an autoclave until nothing, not even a virus, survives. But AIDS has created many other problems in the nation's blood supply: for those, like Dr. Keating and her colleagues, who must find donors and ensure that the blood obtained is safe; for those who give blood; and for those who receive it" (Murray 1990, p. 205).

Scenario 5

In a study of medical vocabulary knowledge among hospital patients, researchers examined the extent to which patients might be failing to understand the meaning of frequently used words. For example, "*Appendectomy* is defined as the 'surgical removal of the appendix vermiformis,' but some respondents hearing it in the statement 'An appendectomy is not serious,' indicated that the term meant a cut rectum, sickness, the stomach, rupture of the appendix, a pain or disease, taking off an arm or leg, something contagious, something like an epidemic, or something to do with the bowels" (Samora, Saunders, Larson 1965, p. 284).

Applied Research

In this chapter we learned that the United States has relied on cheap Mexican labor at least since the 1940s. During the Persian Gulf War, the issue of migrant labor became most apparent when migrant laborers from surrounding countries fled Iraq and Kuwait in the wake of Desert Storm. For this assignment, select one of the major labor-importing countries from the list below and name at least two countries from which they draw labor. In addition, try to find information on a particular occupational category (nurses, hired farmworkers, engineers) that these foreign nationals are known to fill. Major labor-importing countries include

Argentina
Australia
Austria
Belgium
Canada
France
Iraq
Ivory Coast
Kuwait
Libya
Malaysia
Netherlands
New Zealand
Nigeria
Peru
Saudi Arabia
South Africa
Sweden
Switzerland
United Arab Emirates
United Kingdom
Germany
Venezuela

Practice Test

Multiple-Choice Questions

1. There are approximately 2,800 foreign-owned manufacturing plants in Mexico. Most of these are headquartered in
   a. Japan.
   b. Europe.

c. the United States.
d. Korea.

2. _______are sets of principles and definitions about how societies operate and how people relate to one another.
a. Facts
b. Sociological theories
c. Concepts
d. Information

3. __________ is the largest private employer in Mexico.
a. RCA
b. Ford Motor Company
c. Toyota
d. General Motors

4. A conflict theorist would ask which one of the following questions about maquiladoras?
a. Why do maquiladoras exist, and what consequences do they have for the United States and Mexico?
b. Who benefits and who loses under the maquiladora arrangement?
c. Does everyone in the United States and Mexico see maquiladoras in the same way?

5. _______ was the first American company to open a maquila plant.
a. General Motors
b. RCA
c. Johnson & Johnson
d. Baldwin Piano

6. The early functionalists used ______ to illustrate the concepts of function and interdependence.
a. the changing seasons
b. a cloth-weaving analogy
c. the metaphor of a machine
d. biological analogies

7. Sociologist Herbert Gans maintained that the poor serve many functions for society. Which one of the following is not one of those functions?
   a. Most give up looking for work and thus are not counted as unemployed.
   b. The poor do the most undesirable jobs in society at a low wage.
   c. The poor do time-consuming work for affluent persons.
   d. Certain middle-class occupations exist to serve the poor.

8. Critics of the functionalist perspective argue that functionalists are conservatives who justify the way things are. This criticism can be countered with the argument that
   a. certain "parts," no matter how problematic, must exist if the system is to operate smoothly.
   b. functionalists are simply illustrating why certain "parts" continue to exist despite efforts to change or eliminate them.
   c. "parts" are interdependent.
   d. if a part exists, no matter how negative, at some level it must benefit everyone in the society.

9. ______ are consequences disruptive to the system or to some segment of society.
   a. Functions
   b. Dysfunctions
   c. Latent functions
   d. Manifest functions

10. Which of the following is a latent function of the 1986 Immigration Reform and Control Act (IRCA)?
    a. It reduced the number of illegal immigrants entering the United States.
    b. It discouraged American employers from hiring undocumented workers.
    c. When applicants filed for legal status, they received new Social Security accounts, and past payments to other accounts were transferred accordingly.
    d. Employers had difficulty distinguishing between fake and legal documents.

11. Conflict theorists are inspired by
    a. Max Weber.
    b. Emile Durkheim.
    c. Karl Marx.
    d. C. Wright Mills.

12. According to Marx, the bourgeoisie
    a. own only their labor.
    b. search for ways to make production more cost-efficient.
    c. strive to find the most skilled workers.
    d. offer workers economic incentives to increase output.

13. Which one of the questions listed below is a functionalist most likely to ask?
    a. How is social order possible?
    b. How do meanings change over time?
    c. How does a part contribute to societal stability?
    d. Who benefits from a particular pattern or social arrangement, and at whose expense?

14. The _______ is an argument used to justify exploitive practices.
    a. facade of legitimacy
    b. status quo
    c. legitimate lie
    d. justifiable facade

15. A conflict theorist would argue that laid-off factory workers who have trouble finding work
    a. are incapable of meeting entry-level requirements needed to qualify for available jobs.
    b. are functionally illiterate.
    c. function at the fifth- or sixth-grade level.
    d. did mindless, repetitive work that left them with no skills even after decades of employment.

16. The ________ own the means of production.
    a. bourgeoisie
    b. proletariat
    c. middle class
    d. people

17. According to Marx, the proletariat
    a. have considerable leverage over employers because they can always threaten to strike.
    b. are workers who own nothing of the production process except their labor.

c. are in a position to negotiate a decent wage.
d. own nothing of the means of production.

18. A major criticism of conflict theory is that it
    a. overemphasizies the stability and order that exist in a society.
    b. offers a simplistic view of the employer-employee relationship.
    c. focuses too strongly on consumer groups, citizen groups, and workers' ability to promote change.
    d. understates the tensions and divisions that exist in society.

19. Which one of the following topics would a symbolic interactionist choose if asked to write an article on the Persian Gulf War (Desert Storm)?
    a. Class differences between military personnel and government leaders
    b. American perceptions of Saddam Hussein before, during, and after the war
    c. The unifying effect of the Gulf War
    d. The conflict over oil, a scarce and valued resource

20. Symbolic interactionists have drawn much of their inspiration from
    a. C. Wright Mills.
    b. Harriet Martineau.
    c. Max Weber.
    d. George Herbert Mead.

21. Symbolic interactionists believe that during interaction, the parties involved
    a. respond directly to the surroundings and to each other's actions.
    b. make interpretations and then respond accordingly.
    c. usually communicate effectively even if they do not share the same symbol system.
    d. do not have to share a symbol system.

22. One shortcoming of the symbolic interactionist perspective is that it
    a. emphasizes order and stability in society.
    b. offers no systematic framework to guide analysis.
    c. exaggerates the amount of conflict and division that exist in a society.
    d. focuses on how people experience and understand the world.

23. An advertiser searching for a strategy to counteract negative images of Mexico would be most likely to adopt a ________ perspective.
    a. functionalist
    b. conflict
    c. symbolic interactionist

For questions 24-30 decide whether each statement about the *maquila* would be made by a
    a. functionalist.
    b. conflict theorist.
    c. symbolic interactionist.

24. U.S. companies, such as GM, moved labor-intensive assembly operations to countries such as Mexico in order to survive in an atmosphere of intensive global competition.

25. Americans tend to associate Mexico with the afternoon *siesta.*

26. The real purpose of the *maquila* industry is to increase profits by exploiting the most vulnerable and least expensive labor.

27. Advertisers employed a number of strategies to offset the negative stereotypes of Mexicans that Americans tend to hold.

28. American companies gain control over workers and community demands by threatening to relocate to Mexico.

29. Whenever an event alters the way in which a significant number of people earn their livelihoods, we can anticipate that some adjustments will have to be made.

30. *Maquila* jobs offer women an income that increases their status in Mexican society.

31. The Mexican economy is about ______ percent the size of the U.S. economy.
    a. 80
    b. 50
    c. 20
    d. 5

True/False Questions

T F 1. *Braceros* is the name for foreign-owned plants that purchase Mexican labor.

T F 2. In the United States, a suntan has always symbolized working-class status.

T F 3. The symbolic interactionist perspective can be used to support the argument that Mexico is a drain on the American economy.

T F 4. Although Mexican workers earn less than their U.S. counterparts, the time (in hours) required to earn products such as black beans, shampoo, and dark rum is equivalent.

T F 5. The largely negative symbolic associations associated with the color black are related to a universal fear of the dark.

T F 6. There was no real need for the automobile, the photocopier, or the airplane when they were first designed.

T F 7. A true functionalist would argue that necessity is the mother of invention.

T F 8. The phrase "integration with separation" captures the way of life along the U.S.-Mexico border.

T F 9. Men have always been the preferred workers for the unskilled, low-wage jobs that characterize the *maquiladoras*.

T F 10. No single theoretical perspective can give us a complete picture of social events.

T F 11. Another name for the *maquiladora* program is NAFTA.

Continuing Education

Leisure

For a change in music, listen to:

Los Grandes del Folklore Mexicano (various artists). RCA International.
Masses and Festival Music of Mexico. Arion.
The Real Mexico (Henrietta Yurchenco). Elektra/Nonesuch.

The next time you rent a movie, consider

The Gods Must Be Crazy, a South African film directed by Jamie Uys (109 minutes). In this movie a Coke bottle drops from a plane and is found by a Kalahari Desert Bushman. Until this point the Kalahari have been an extremely peaceful and loving people. The bottle changes their way of life and their relationships with one another. As you view the film, consider which parts a functionalist, a conflict theorist, and a symbolic interactionist might emphasize. A functionalist would ask what functions and dysfunctions (latent and manifest) the Coke bottle had for the Kalahari. A conflict theorist would examine themes of conflict over a scarce and valued resource (the bottle). A symbolic interactionist would focus on the meanings the Kalahari assigned to the bottle.

Travel

If you're in Detroit, visit the Detroit Institute of Arts (5200 Woodward Avenue; 313-833-7900). In particular, see the murals by famous Mexican artist Diego Rivera, depicting Detroit industry.

Look for paintings by the Mexican artist Frida Kahlo, a woman who lived in Diego Rivera's shadow (and married him twice). One way to learn about this artist is through the film *Frida Kahlo: A Ribbon around a Bomb*, directed by Ken Mandel. Madonna, Robert de Niro, Luis Valdez, and Luis Mandoki are all working on film projects dealing with Kahlo's life and art (Zaniello 1992).

If you're in San Antonio, visit El Mercado/The Mexican Market (W. Commerce Street and Santa Rosa Street; 512-299-8600). The market, located in the heart of the

Mexican quarter, has a lively atmosphere. Folk art, handicrafts, and Mexican foodstuffs may be purchased there.

General and Readable Information

Twin Plant News: The Magazine of the Maquiladora and Mexican Industries (4110 Rio Bravo Drive, #108, El Paso, Tx 79902) is a monthly magazine that covers all aspects of offshore manufacturing in Mexico. Some recent features include "*Maquiladora* Industry the Economic Impact on San Diego's Economy," "Border Waste Program...When Two Cultures Meet," and "Maquila Concept Now Worldwide." The magazine is pro-*maquiladora*, and it promotes conscientious business practices. If you're looking for a highly critical overview of the *maquiladora* industry, this is not the journal to read.

Grassroots Development: Journal of the Inter-American Foundation (901 N. Stuart Street, 10th Floor, Arlington, VA 22203) is free and it is published three times a year in English, Spanish, and Portuguese. Its purpose is to explore how development assistance can contribute more effectively to self-help efforts. The journal reports on how the poor in Latin America and the Caribbean organize and work to improve their lives.

Voices of Mexico: Mexican Perspectives on Contemporary Issues (Miguel Angel de Quevedo 610, Col. Coyoacan, 04000 Mexico, D.F.) is published bi-monthly. It offers "in-depth and up-to-date coverage in English of Mexico's social, cultural, political and economic life." Each issue includes "analysis of vital issues, as well as thought-provoking opinion and debate."

Country Profile
Mexico

Official Name: United Mexican States
Population (1994 est.): 91.8 million
Land Area: 764,000 square miles
Population Density: 125 people per square mile
Birth Rate per 1,000: 28

Death Rate per 1,000: 6
Rate of Natural Increase: 2.2%
Average Number of Children Born to a Woman During Her Lifetime: 3.2
Infant Mortality Rate: 35 deaths per 1,000 live births
Life Expectancy at Birth: 70 years
Gross National Product per Capita: $3,470
Capital: Mexico City

Mexico is the most populous Spanish-speaking country in the world; it is the second most populous country in Latin America, after Portuguese-speaking Brazil. About 70% of the people live in urban areas. Many Mexicans emigrate from rural areas that lack job opportunities--such as the underdeveloped southern states and the crowded central plataeau--to the industrialized urban centers and the developing areas along the US -Mexico border. According to some estimates, the population of the areas surrounding Mexico City is about 20 million, which would make it the largest concentration of population in the world. Cities bordering on the United States such as Tijuana and Ciudad Juarez, and cities in the interior such as Guadalajara, Monterrey, and Puebla, have undergone sharp rises in population.

Education in Mexico is being decentralized and enhanced in rural areas. The increase in school enrollments during the past two decades has been dramatic. Education is mandatory from ages six through 18. Primary enrollment from 1970 through 1993 increased from less than 10 million to 15 million. In 1993, 59% of the population between the ages of six and 18 were enrolled in school. Enrollments at the secondary school level also shot up from 1.4 million in 1972 to as many as 4 million in 1993. This rapid rise occurred in higher education also. Between 1959 and 1993, college-level enrollments rose from 62,000 to 1.2 million. Education spending has risen dramatically from 2.6% of GDP in 1988 to 4% in 1993. The 1994 education budget is 4.4% of GDP.

Contemporary artists, architects, writers, musicians, and dancers draw inspiration from a rich history of Indian civilization, colonial influence, revolution, and the development of the modern Mexican state. Artists and intellectuals alike emphasize the problems of social relations in a context of national and revolutionary traditions.

Sources: Haub and Yanagishita (1994); U.S. Department of State (1994).

Chapter References

Haub, Carl and Machiko Yanagishita. 1994. *1994 World Population Data Sheet.* Washington, DC: Population Reference Bureau.

Light, Ivan and Edna Bonacich. 1988. *Immigrant Entrepreneurs: Koreans in Los Angeles 1965-1982.* Los Angeles: University of California Press.

Murray, Thomas H. 1990. "The Poisoned Gift: AIDS and Blood." *The Milbank Quarterly* 68(2):205-25.

Ogle, George E. 1990. *South Korea: Dissent Within the Economic Miracle.* Atlantic Highlands,NJ: Zed Books.

Rotella, Sebastian. 1994. "Tijuana Battles for Respect." *Los Angeles Times* (April 22):A1+.

Samora, Julian, Lyle Saunders, and Richard F. Larson. 1965. "Medical Vocabulary Knowledge Among Hospital Patients." Pp. 278-91 in *Social Interaction and Patient Care* edited by J. K. Skipper, Jr. and R. C. Leonard. Philadelphia: Lippincott.

U.S. Department of State. 1994. "Mexico." *Background Notes* (#7365). Washington, DC: U.S. Government Printing Office.

Zaniello, Tom. 1992. *The Real Movies Calendar.* (February 1).

Answers

Concept Application

1. Functions
2. Facade of legitimacy; Ideology
3. Symbols
4. Latent dysfunctions
5. Symbols

Multiple-Choice

1.c 2.b 3.d 4.b 5.b 6.d 7.a 8.b 9.b 10.c 11.c 12.b 13.a 14.a 15.d 16.a 17.b 18.b 19.b 20.d 21.b 22.b 23.c 24.a 25.c 26.b 27.c 28.b 29.a 30.a 31.d

True and False

1.F 2.F 3.F 4.F 5.F 6.T 7.T 8.T 9.F 10.T 11.F

Chapter 3
Research Methods

Study Questions

1. Define research methods.

2. Distinguish between data and information.

3. Why does Japan receive special attention in a chapter on research methods?

4. What is the formula typically used to calculate the trade deficit between two countries? State the major shortcomings of this formula.

5. What is the advantage of understanding the methods of social research?

6. What is the information explosion? What technological innovations are responsible for this phenomenon? Explain.

7. What factors does Orrin Klapp identify as the causes underlying distorted, exaggerated presentation of data?

8. What is meant by *research methods literate*?

9. What assumptions underlie the scientific method? Under what circumstances do research findings endure?

10. Why did Dean Barnlund conduct the research for *Communicative Styles of Japanese and Americans: Images and Realities*?

11. Why is it important for researchers to explain their reasons for investigating a particular topic?

12. Why did researcher Anne Allison choose to study the obento? What is the significance of the obento for preschool education?

13. Why should researchers review the literature before beginning to investigate a topic? What are some general strategies for conducting a literature review?

14. What distinguishes one discipline from another? Expand on your answer using suicide as an example.

15. What is a variable? Give an example. What is the difference between an independent and a dependent variable?

16. What is a hypothesis?

17. Identify the concepts, the variables (independent and dependent), and at least one hypothesis that Barnlund used in his research.

18. Identify and briefly describe the units of analysis that sociologists study.

19. Distinguish between a population and a sample.

20. What is a random sample? Why is it difficult to secure a random sample?

21. Give a brief description of each method of data collection, and state what you believe might be the disadvantages of each method.

Self-Administered
Questionnaires

Structured
Interviews

Unstructured
Interviews

Participant
Observation

NonParticipant
Observation

Secondary
Sources

22. Name two adjustments that researchers Raymond A. Jussaume, Jr. and Yoshiharu Yamada made to increase the chances that American and Japanese respondents would react favorably to their request to participate in a research project. State the rationale for each adjustment.

23. What is an operational definition? What qualities make an operational definition useful as a data-gathering tool?

24. Specify Barnlund's unit of analysis, sample, and method of data collection, and his operational definition of a stranger. Comment on the reliability and validity of this operational definition.

25. How do social researchers go about analyzing their data?

26. What meaningful patterns did Barnlund find after he analyzed Japanese and American students' responses to his questions?

27. When can findings be considered generalizable to the larger population?

28. What three conditions must be met before a researcher can claim that an independent variable contributes significantly to explaining a dependent variable?

29. What is a probabilistic model of cause?

30. Discuss the generalizability of Barnlund's findings. Can we say with certainty that culture is the variable which explains the differences between Japanese and American behavior toward strangers?

Concept Application

Below are five scenarios and the sources from which they were drawn. Decide which concept or concepts covered in Chapter 3 is represented best by each scenario, and explain why. The following concepts are considered:

Concepts
Control variables
Correlation
Dearth of feedback
Dependent variable
Documents
Generalizability
Household
Hypothesis
Independent variable
Information explosion
Observation
Participant observation
Reliability
Secondary sources
Self-administered questionnaire
Spurious correlation
Structured interview
Traces

Scenario 1

"With the appearance of the tape-recorder, a monster with the appetite of a tapeworm, we now have a new problem of what I call artificial survival. The effort needed to write a book, even of memoirs, requires discipline and perseverance which until now imposed a certain natural selection on what survived in print. But with all sorts of people being encouraged to ramble effortlessly and endlessly into a tape-recorder...some veins of gold and a vast mass of trivia are being preserved which would otherwise have gone to dust " (Tuchman 1982, p. 72).

Information Explosion

Scenario 2

"Acceptable/appropriate attitudes for drinking alcoholic beverages were assessed by asking, 'How much should a man/woman of your age feel free to drink in the following situations:

Operational definition ~~observation~~

- ø At a bar with friends.
- ø At a party at someone else's house.
- ø As a parent, spending time with small children.
- ø During working hours (not just at lunch).
- ø Visiting in-laws.
- ø With friends at home.
- ø Getting together with friends after work before going home.
- ø Getting together with people at sports events or recreation.
- ø Going to drive a car.'

Respondents were asked to choose one of the following four categories of response: (1) no drinking, (2) drinking only small amounts, (3) drinking enough to feel effects but not enough to get drunk and (4) getting drunk is sometimes all right" (Tsunoda et al. 1992, p. 370).

Scenario 3

"The popularity of a President is usually predicted quite closely by people's sense of where the economy is going. [Data shows that when] people sense that the economy is probably improving, they tend to approve of the President. When they sense that the economy's declining, they tend to disapprove of the President" (Chomsky 1989, p. 40).

Correlation

Scenario 4

"[A] major question was whether lead could poison a child even at levels that could ordinarily be found in older homes because of the shedding of lead-based paint. The levels measured in children's blood were difficult to assess because they fluctuated week to week depending on exposures. But how much cumulative damage might be done?

Dr. Needleman found a clever approach to the question when he went to first- and

operational Definition Traces Correlation

second-grade classes in Boston, and passed out small rewards for the children to bring in their baby teeth as they fell out. Lead builds up steadily in teeth.

Measuring the amount of lead in slices of each tooth, and then comparing the children with the lowest and highest levels of lead, he found that those with the most exposure had lower I.Q. scores by an average of 3 or 4 points" (Hilts 1992, p. D28).

operational def., Traces, Correlation

Scenario 5

"The two of us began to collect raw data. We took care not to ask questions about time or even to appear to be interested in time. We simply recorded what people said and did in the course of their everyday lives that had anything to do with time. For instance, a colleague poked his head around our door and said, 'I'll be gone for a while.' We would record 'a while' and question 'How long is a while?' Noting when he returned, we saw that 'a while' meant more than the quarter or half an hour that was usual to be away from one's desk " (Hall 1992, p. 223).

operational definition observation

Applied Research

Find a research article in which the authors study behavior in at least two countries. Give a brief description of the following elements:

Topic
Major concepts
Independent and dependent variables (if applicable)
Unit of analysis
Methods of data collection
Operational definition (if applicable)
Findings
Conclusion

Practice Test

Multiple-Choice Questions

1. The unprecedented rate of increase in the volume of information due to the development of the computer and telecommunications is
   a. the computer revolution.
   b. megatrends.
   c. the information explosion.
   d. the telecommunications revolution.

2. Japan receives special attention in the chapter on research methods for which one of the following reasons?
   a. A tremendous amount has been written about Japan, especially about its rise from the ashes of World War II to become a major economic power.
   b. Japan is the only country in the world with which the United States has a trade deficit.
   c. Japan is a resource-rich country.
   d. U.S. sociologists do a considerable amount of research comparing Japanese and American societies.

3. The most accurate operational definition of the trade balance between two countries would consider all but which one of the following items?
   a. The dollar value of goods and services that each country exports to the other.
   b. The dollar value of what each country produces and sells within the other.
   c. The dollar value of what each country exports indirectly to the other.
   d. The dollar value of the products that are barred from entering each country.

4. Which analogy did sociologist Orrin Klapp use to describe the dilemma of sorting through and keeping up with the massive amount of information being generated?
   a. A sociologist drowning in quicksand
   b. A student trying to take notes while 10 professors talk at one time
   c. A researcher working on a gigantic jigsaw puzzle while additional pieces are flowing onto the table from a funnel overhead
   d. A person entering a crowded six-lane highway with thousands of signs

5. Klapp used this analogy (see Question 4) to show that
   a. most people cannot possibly register all the messages they encounter during the day.
   b. the speed by which information is produced and distributed overwhelms the brain's capacity to organize and evaluate it.
   c. there are thousands of important newspapers, magazines, and journals that people will never have the time to read.
   d. people do not have the reading and writing skills to comprehend information.

6. Klapp maintains that in the context of the information explosion, poor-quality information exists because
   a. no one pays attention.
   b. there is a dearth of feedback.
   c. the creators of this information lack basic research skills.
   d. few people are computer literate.

7. Which one of the following assumptions applies to the scientific method?
   a. Knowledge is always subjective.
   b. Research findings can be manipulated to advance a good cause.
   c. Truth is confirmed through faith.
   d. Knowledge is acquired through observation.

8. Perhaps one of the most significant and most often underestimated reason why a researcher chooses to study a specific topic is
   a. that funding is available.
   b. sociological appeal.
   c. personal interest.
   d. to understand how society works.

9. One of the least important criteria for selecting literature to review is
   a. that it represents an opposing view.
   b. the publication date.
   c. the publisher's reputation.
   d. the author's reputation.

10. ______ are powerful thinking and communication tools that enable people to give and receive complex information in an efficient manner.
   a. Variables

b. Hypotheses
c. Concepts
d. Measures

11. Sociologists use the concept of suicide to mean
a. the severing of relationships.
b. an act of self-destruction.
c. an intentional act of aggression turned inward.
d. a chemical imbalance that leads to a destructive act.

Use the following hypothesis to answer questions 12 and 13:
Japanese are less likely than Americans to notice strangers.

12. The independent variable in the hypothesis is
a. Japanese culture.
b. American culture.
c. cultural background.
d. the tendency to notice strangers.

13. The dependent variable in the hypothesis is
a. Japanese culture.
b. American culture.
c. cultural background.
d. the tendency to notice strangers.

Use the following hypothesis to answer questions 14 and 15:
Rural populations have a higher suicide rate than urban populations in Japan.

14. The independent variable in the hypothesis is
a. rural populations.
b. urban populations.
c. geographical location.
d. the suicide rate.

15. The dependent variable in the hypothesis is
a. urban populations.
b. Japan.

c. culture.
d. the suicide rate.

16. _____ are settings that have borders or that are set aside for particular activities.
   a. Traces
   b. Territories
   c. Households
   d. Documents

17. ______ are the unit of analysis for a research project that collects garbage from landfills and selected neighborhood garbage cans.
   a. Traces
   b. Documents
   c. Territories
   d. Households

18. ______ are the unit of analysis for a research project in which the researcher visits and observes cities to learn how residents use urban space.
   a. Traces
   b. Documents
   c. Territories
   d. Households

19. A sampling frame is
   a. a complete list of every case in a population.
   b. a portion of cases from a particular population.
   c. the plan for gathering data to test hypotheses.
   d. a sample with the same distribution of characteristics as the population from which it is drawn.

20. In his study of behavior toward strangers, Barnlund studied
   a. a random sample of Japanese and Americans living in Tokyo and in New York respectively.
   b. Japanese and American businessmen and businesswomen employed in large corporations.

c. Japanese and American college students enrolled in public and private universities in four cities in each country.

d. Japanese and Americans who work for companies that have offices in both countries.

Use one from the following set of responses to answer questions 21-25:

a. Self-administered questionnaire
b. Structured interview
c. Unstructured interview
d. Nonparticipant observation
e. Participant observation
f. Secondary analysis

Below are excerpts from reviews of research published in *Contemporary Sociology*. On the basis of each description, determine the one method of collecting data.

21. The author of *Gender Concepts of Swedish and American Youth* distributed questionnaires "to both Swedish and American eleven-, fourteen-, and eighteen-year olds. Subjects were asked to list attributes they thought characterized most women, most men, most boys, most girls, and themselves; to write 'change-sex stories,' about what their lives would be like if they found they had become the other sex; to complete the Bem Sex Role Inventory and other commonly used measures of gender constructs; and to report their life expectations" (Howard 1989, p. 341).

22. "*Ambiguous Lives* is a reconstruction of the lives of free women of color, a subcaste within the African-American community, in antebellum Georgia. The author, Adele Logan Alexander, uses her own family to typify that subcaste. Employing a variety of sources--oral history; archival information; legal documentation; newspapers; federal government reports; school, church, and community records--she weaves a complex web of racial dynamics, heretofore untold, of free women of color and their families, who were isolated in the rural middle-Georgia plantation belt before Reconstruction " (Benjamin 1992, p. 454).

23. *Fire from the Mountain*, by Omar Cabezas, is an account of the author's experience as a guerrilla for the FSLN (Sandinista National Liberation Front). Cabezas is honest,

and he captures the pain and commitment that sculpted the revolution. He loses his close friend and teacher in the mountains, and agonizes with self-doubt and loneliness; he cannot visit his mother while underground in his hometown; his companera leaves him; and he cannot visit his new daughter. He delivers all this without martyrdom or bravado, but with disarming candor that frequently sees the humor in his own situation. It is this humor that most effectively demythologizes, allowing us to see a real, 'unofficial' history, and not a manufactured one" (Legeay 1987, p. 353).

24. "The main message of Mark Stern's comparison of the civil rights policies of John Kennedy and Lyndon Johnson is conveyed by the chapter titles. For Kennedy, we get 'An Intimidated President' and 'A Reluctant Commitment'; for Lyndon Johnson, we get 'The Second Great Emanicipator.' Kennedy, the book tells us, received more credit for supporting civil rights than he deserved, Johnson far less. Based on extensive archival research, the book covers the development of each man's thinking on racial issues (giving more space to Johnson) and describes the evolution of each man's civil rights politics in the context of their respective political styles. On the latter, for example, we learn that Kennedy hardly ever asked a member of Congress to vote a certain way on an issue; a sharper contrast with Johnson is hard to imagine" (Payne 1993, p. 73).

25. In *Waiting for Disaster,* researchers sat down one-on-one with people affected by an earthquake on New Year's Day in 1979. "Only about one-third in the affected area reported having been 'very frightened or somewhat frightened,' and only one person in eight made an effort to contact someone either for information or because of concern for another person's welfare. Yet two-thirds remained 'very frightened' or 'somewhat frightened' at the prospect of a damaging earthquake in the near future" (Lang, 1987, p. 785).

26. Which of the following is one of the ways in which researchers Jussaume and Yamada adjusted their questionnaire to accommodate Japanese respondents?
    a. Sent the questionnaire in an American-style envelope to arouse the Japanese respondents' curosity.
    b. Signed each letter to give it a personal touch.
    c. Handwrote the address.
    d. Mentioned that the purpose of the research was to learn about bilateral trade with the United States.

27. It takes one person counting aloud approximately _________ to count to one billion.
    a. 3 years
    b. 32 days
    c. 3 days
    d. 32 years

28. Which one of the following time periods would generate the most reliable answer to "During the past ________, how often did you notice and think about a stranger's personal characteristic?"
    a. month
    b. year
    c. week
    d. 24 hours

29. Upon analyzing the data, Barnlund found that when interaction takes place with a stranger, the Japanese most frequently learn the stranger's __________; while Americans most frequently learn the stranger's __________.
    a. address; name
    b. occupation; address
    c. occupation; name
    d. address; phone number

30. Under which one of the following conditions are findings from a sample not generalizable to a larger population?
    a. The sample is a series of interesting case studies.
    b. The response rate is high.
    c. Almost all the subjects agree to participate.
    d. The sample is random.

31. If a correlation between academic achievement and a physical trait such as skin color disappears in the presence of the variable after-school employment, we know that
    a. skin color is responsible for differences in academic achievement.
    b. after-school employment is not responsible for differences in academic achievement.
    c. the relationship between after-school employment and skin color is spurious.
    d. academic achievement should be treated as a control variable.

32. Which one of the following conditions need not be met by researchers before they can claim that an independent variable contributes significantly to explaining a dependent variable?
   a. The independent variable must precede the dependent variable in time.
   b. The independent and the dependent variables must be correlated.
   c. There must be no evidence that a third variable is responsible for a spurious correlation between the independent and the dependent variables.
   d. The independent variable cannot be an ascribed characteristic.

33. If we know that the total number of phone messages transmitted between Japan and the United States is 97 million and that the number of phone messages transmitted between Israel and the United States is 22 million, we can
   a. draw the conclusion that Japan is more interdependent with the United States than is Israel.
   b. draw the conclusion that Israel is more interdependent with the United States than is Japan.
   c. make no assumption about which country is more interdependent with the United States until we convert phone messages to per capita terms.
   d. conclude that phone messages are not an accurate measure of interconnection because we don't know who is making the calls.

True/False Questions

T F 1. A working knowledge of research methods is necessary only if a person plans to do research.

T F 2. The best researchers always study representative samples.

T F 3. In a probabilistic model, the independent variable explains the dependent variable with 100 percent accuracy.

T F 4. The trade deficit is a valid measure of the overall health of the American economy.

T F 5. The trade balance is calculated most accurately by adding up the dollar value of the goods and services that each country exports to the other and then subtracting one number from the other.

T F 6. Research findings endure if they can withstand continued reexamination and duplication by scientists.

T F 7. Researchers manipulate data to support well-intentioned personal, economic, and political agendas.

T F 8. Researchers always follow the six steps of research in sequence.

T F 9. Obtaining a random sample is relatively easy.

T F 10. The trade deficit, as traditionally calculated, is an accurate measure of the openness of the Japanese market to U.S. goods and services.

T F 11. Research suggests that in comparison with American fathers, Japanese fathers spend little or no time with their children.

T F 12. A correlation between two variables is evidence that one variable causes the other.

T F 13. Upon reviewing Barnlund's findings, we can say with certainty that culture explains differences in the ways Japanese and Americans behave toward strangers.

Continuing Education

Leisure

For a change in music, listen to:

Heart of the Wind: Japanese Shakuhachi Music (Masayuki Koga). Fortune.
The Hiroshima Masses. Lyrichord.
UNESCO Collection: O-Suwa-Daiko (Japanese Drums). Grem/Waira.

The next time you rent a movie, consider

Pumping Iron II: The Women. This documentary, an excellent example of nonparticipant observation research, is directed by George Butler (107 minutes). The film offers an in-depth look at the world of female bodybuilding. At the same time, it explores social factors that constrain women from achieving "perfection" in this sport. The subjects studied are clearly not representative of the female population; yet their experiences clarify important issues with regard to constraints on females' efforts to achieve their potential. As you view the film, think about the extent to which the Hawthorne effect may be operating. (The Hawthorne effect is a phenomenon whereby observed persons alter their behavior when they know they are part of a research project.) Identify subjects that you believe are "managing" their behavior for the camera. Do you think we can acquire important information even when behavior is managed?

Travel

If you're in Boston, visit the Computer Museum (300 Congress Street; 617-423-6758). The various data processing exhibits provide an excellent overview of the technology behind the information explosion.

If you're in Bozeman, Montana, visit the American Computer Museum (234 E. Babcock Street; 406-587-7545). According to *The New Yorker*, "The museum is arranged chronologically. It begins with paintings that show Egyptian hieroglyphics and Roman and Hindu-Arabic numerals and ends with a robotic arm that picks up wooden alphabet blocks and sets them in a row, to spell 'ROBOT'" (1991, p. 32).

General and Readable Information

Japan Quarterly (5-3-2 Tsukiji, Chuo-ku, Tokyo 104-11, Japan) is published four times a year in English. It features articles, critiques, reviews, and synopses of recent publications on all aspects of contemporary Japan, including politics, foreign affairs, finance, culture, technology, literature, the arts, women's issues, social trends, and folklore.

Country Profile
Japan

Official Name: Japan
Population (1994 est.): 125.0 million
Land Area: 145,370 square miles
Population Density: 860 people per square mile
Birth Rate per 1,000: 10
Death Rate per 1,000: 7
Rate of Natural Increase: 0.3%
Average Number of Children born to a Woman During Her Lifetime: 1.5
Infant Mortality Rate: 4 deaths per 1,000 live births
Life Expectancy at Birth: 79 years
Gross National Product: $28,220 per Capita
Capital: Tokyo

Japan is one of the most densely populated nations in the world, with almost 318 persons per square kilometer (823 persons per sq. mi.). The growth rate has stabilized at about 0.5% in recent years, giving rise to some concern about the social implications of an increasingly aged population.

The Japanese are a Mongoloid people, closely related to the major groups of East Asia. However, some evidence also exists of admixture with Malayan and Caucasoid strains. About 675,000 Koreans and much smaller groups of Chinese and Caucasians reside in Japan.

Buddhism is important in Japan's religious life and has strongly influenced fine arts, social institutions, and thought. Most Japanese still consider themselves members of one of the major Buddhist sects.

Shintoism is an indigenous religion founded on myths, legends, and ritual practices of the early Japanese. Neither Buddhism nor Shintoism is an exclusive religion; most Japanese observe both Buddhist and Shinto rituals, the former for funerals and the latter for births, marriages, and other occasions. Confucianism, more an ethical system than a religion, profoundly influences Japanese thought.

About 1.5 million people in Japan are Christians; approximately 60 percent are Protestant and 40 percent Roman Catholic.

Japan provides free public schooling for all children through junior high school. Ninety-four percent of students go on to three-year senior high schools, and competition is fierce for entry into the best universities. Students may attend either public or private high schools, colleges, and universities, but they must pay tuition. Japan enjoys one of the world's highest literacy rates (99%); nearly 90% of Japanese students complete high school.

Sources: Haub and Yanagishita (1994); U.S. Department of State (1990).

Chapter References

Benjamin, Lois. 1992. Review of *Ambiguous Lives: Free Women of Color in Rural Georgia, 1789-1879*, by Adele Logan Alexander. *Contemporary Sociology* 21(4):454-55.

Chomsky, Noam. 1989. "Noam Chomsky, Linguist." Pp. 38-58 in *A World of Ideas: Conversations with Thoughtful Men and Women About American Life Today and the Ideas Shaping Our Future* by Bill Moyers. New York:Doubleday.

Hall, Edward T. 1992. *An Anthropology of Everyday Life: An Autobiography*. New York: Doubleday.

Haub, Carl and Machiko Yanagishita. *1994. 1994 World Population Data Sheet.* Washington, DC:Population Reference Bureau.

Hilts, Philip J. 1992. "Hearing Is Held on Scientist's Lead-Poison Data." *The New York Times* (April 15):D28.

Howard, Judith A. 1989. Review of *Gender Concepts of Swedish and American Youth* by Margaret Jean Intons-Peterson. *Contemporary Sociology* 18(3):340-42.

Lang, Gladys Engel. 1987. Review of *Waiting for Disaster: Earthquake Watch in California*, by Ralph H. Turner, Joanne M. Nigg, and Denise Heller Paz. *Contemporary Sociology* 16(6):784-87.

Legeay, Stephen P. 1987. Review of *Fire from the Mountain: The Making of a Sandinista,* by Omar Cabezas. *Contemporary Sociology* 16(3):352-53.

The New Yorker. 1991. "The Sweep of Time." (October 7):30-32.

Payne, Charles M. 1993. Review of *Calculating Visions: Kennedy, Johnson, and Civil Rights*, by Mark Stern. *Contemporary Sociology* 22(1):73-74.

Tsunoda, Tooru, Kiyoko M. Parrish, Susumu Higuchi, Frederick S. Stinson, Hiroaki Kono, Motoi Ogata, and Thomas C. Harford. 1992. "The Effect of Acculturation on

Drinking Attitudes among Japanese in Japan and Japanese Americans in Hawaii and California." *Journal of Studies on Alcohol* 53(4):369-77.

Tuchman, Barbara W. 1982. *Practicing History: Selected Essays.* New York: Ballantine.

U.S. Department of State. 1990. "Japan." *Background Notes* (#7770). Washington, DC: U.S. Government Printing Office.

Answers

Concept Application

1. Information explosion
2. Operational definition
3. Correlation
4. Operational definition; Traces; Correlation
5. Operational definition; Observation

Multiple-Choice

1.c 2.a 3.d 4.c 5.b 6.b 7.d 8.c 9.b 10.c 11.a 12.c 13.d 14.c 15.d 16.b 17.a 18.c 19.a 20.c 21.a 22.f 23.e 24.f 25.b 26.c 27.d 28.d 29.c 30.a 31.c 32.d 33.c

True/False

1.F 2.F 3.F 4.F 5.F 6.T 7.F 8.F 9.F 10.F 11.F 12.F 13.F

Chapter 4
Culture

Study Questions

1. Why was South Korea chosen to illustrate the material on culture?

2. Is there a clear-cut distinction between Eastern and Western cultures? Why or why not? Explain.

3. What is material culture? What qualities of material culture do sociologists find significant?

4. What is nonmaterial culture?

5. Distinguish between beliefs and values.

6. What are norms? Distinguish between folkways and mores.

7. What is the connection between culture and geographical and historical forces?

8. Explain: "Babies are destined to learn the ways of the culture into which they are born and raised."

9. What do sociologists mean when they say that language is a thinking tool?

10. What are some of the ways in which culture stimulates and satisfies appetite?

11. What makes an emotion social?

12. What are feeling rules?

13. What is the relationship between material and nonmaterial cultures? Give an example.

14. What is diffusion? Give two examples of the diffusion process. Is this process indiscriminate?

15. How does the concept of diffusion relate to issues in international trade?

16. What is culture shock? How is it related to ethnocentrism?

17. What are the various types of ethnocentrism? Give examples of each. (Don't forget reverse ethnocentrism.)

18. What perspectives should one follow when studying other cultures?

19. Does everyone who lives in a particular country share the same culture? What concepts do sociologists use to distinguish groups who depart from the so-called mainstream culture?

20. How would a functionalist, a conflict theorist, and a symbolic interactionist view culture?

21. According to Edward T. Hall, what behaviors must Americans adopt if they are to be effective in communicating and interacting with people from other countries?

Concept Application

Below are five scenarios and the sources from which they were drawn. Decide which concept or concepts covered in Chapter 4 are represented best by each scenario, and explain why. The following concepts are considered:

Beliefs
Counterculture
Cultural genocide
Cultural relativism
Culture shock
Diffusion
Ethnocentrism
Feeling rules
Folkways
Institutionally complete
Material culture
Mores
Nonmaterial culture

Norms
Reverse ethnocentrism
Social emotions
Society
Subcultures
Technology
Values

Scenario 1

"Streets of gold: that is the vision they carry setting out from their villages heading to *el norte* [the north]. There are wonderful tales about the other side: the good jobs, the beauty...and the treasures! People bring back televisions, VCRs, and these new microwaves that cook dinner in less than a minute! Many an afternoon has been spent...going over and over those stories, stories the younger generation are raised on " (Davis 1990, p. 171).

Scenario 2

"Yesterday, my 4-year-old stopped crying. He fell off his bike, held his breath and gritted his teeth. 'I'm not gonna cry, Mom,' he said. 'I'm really not.'

Where did this pint-size stoicism come from? Batman videos? Preschool name-callers? Maybe the neighbors who tell their kid, 'Crying will get you nowhere.' You hear it everywhere: You'd better not pout, you'd better not cry. Big girls don't cry. Grin and bear it, hide it, stifle it, but whatever you do, don't cry, *please*, don't cry. I'll give you a cookie if you stop" (Hogan 1994, p. E1).

Scenario 3

"Samuel Phillips Verner had come to the Congo on a mission, set by the organizers of the Saint Louis World's Fair. The organizers aimed to dazzle fairgoers with a definitive display of Darwinism. To that end, Verner was looking for Pygmies, and he came across one in the person of Ota Benga.

There were also special agents in charge of getting Ainu, Japanese, Filipinos, South Americans, native Americans and lots of North American Indians, including Geronimo. The point is that in memory of the Saint Louis Fair, what's remembered, if anything, is the song Judy Garland [sang] in the movie 'Meet Me in Saint Louis, Louie,' but the real innovation of the Saint Louis Fair was its anthropology department. And what that meant is you took as many people as you could gather from around the world and displayed them as creatures on the evolutionary scale that led up to the caucasian race, who built the Saint Louis Fair, and were obviously the peak of evolution" (*National Public Radio* 1992, p. 36).

Scenario 4

"The peanut is a South American original, a shrub whose flowers send tendrils into the ground where they grow into seed pods. It was domesticated about 4,000 years ago in the eastern foothills of the Andes, somewhere around the border between Bolivia and Argentina. And very thoroughly domesticated; it's one of the few plants in the world that is never found in the wild. By the time of Columbus, it had spread throughout Latin America and the Caribbean.

It was in Africa that the peanut found its heartiest welcome. The Portuguese brought it to West Africa in the early 1500s, and in 1564 the traveler Alvares de Almada reported it was already an established crop in Senegal and Gambia (which are still among the world's greatest peanut-exporting countries). Within 200 years it had spread on its own all the way across Africa to Angola, without Portuguese help" (Perry 1994, p. H9).

Scenario 5

"Japanese frequently bow to one another, for instance when greeting someone, as a gesture of respect and sincerity. The type of bow depends on the formality of the situation, the type of personal relationship (e.g., close or distant), and the differences in social status of the individuals involved. The bow might be no more than a simple nod of the head or, on more formal occasions, a deeper bow from the waist. The most formal bow involves kneeling, placing one's hands out in front on the floor, and lowering the head slowly so that it almost touches the floor. Bowing is not always required, however. Family members and close friends do not usually bow to each other, but a child might bow to his or her mother when apologizing for mischievous behavior" (Japan Information Center 1988, p. 61).

Applied Research

One way to gain a firsthand and personal appreciation for the comparative approach is to talk with an international student attending your college. Once you find out what country the student is from, how he or she came to attend this college, and his or her academic interests, ask the following questions:

1. Before you came to the United States, what did you expect it to be like? How did it differ from your expectations?

2. What do you miss most? (or: What was the biggest adjustment that you had to make, once you arrived in the United States?)

3. What do you like most about life in the United States? Least?

4. Are there any words or expressions in your native language that you cannot find English words for? Can you give some examples? Are there any English words or expressions for which you can find no equivalent words in your native language?

Practice Test

Multiple-Choice Questions

1. In the final analysis, _________ comprises the things people have, the things they do, and what they think.
   a. society
   b. culture

c. material culture
d. beliefs

2. Which one of the following is a society?
   a. Koreatown
   b. Girl Scouts of America
   c. People of Chinese descent
   d. Postal employees

3. Which one of the following is a true statement about the nature of culture?
   a. Culture is a society's distinctive and complete design for living.
   b. Genes and personality are clearly more powerful influences on human behavior than is culture.
   c. Culture is everything that people acquire from other societies.
   d. Culture and physical appearance are intimately related.

4. South Korea is
   a. a minor player in the Asian economic community.
   b. a society with a Japanese-like culture.
   c. a country experiencing tremendous economic decline.
   d. one of 12 countries and city-states in the Pacific Rim.

5. Sociologists use the term technology to mean
   a. all the objects or physical substances available to the people of a society.
   b. intangible human creations.
   c. guidelines that govern routine uses of resources.
   d. the knowledge, skills, and tools used to transform resources into forms with specific purposes, and the skills and knowledge to do so.

6. Which one of the following behaviors tends to be characteristic of Korean culture?
   a. Koreans fill every possible space on a sheet of paper before throwing it away.
   b. When Koreans view an outstanding individual achievement, they tend to give more credit to individual talent than to hard work.
   c. Koreans have clearly marked eating spaces.
   d. Koreans admire Japanese ways and try to model themselves after Japanese.

7. Which one of the following is an example of mores?
   a. It is appropriate to reach across another person's space to take food from the serving bowls.
   b. Open the refrigerator only as far as necessary to remove an item.
   c. Throw away a sheet of paper with one line of writing if you don't like what you wrote.
   d. Do not take the life of another person except for a just cause.

8. Consumption- and conservation-oriented behaviors seem to be related to
   a. genetic qualities.
   b. culture.
   c. issues of resource abundance and scarcity.
   d. population size.

9. In the United States the word *individual* connotes
   a. interdependence.
   b. intertwinement.
   c. one of a kind.
   d. selflessness.

10. One indicator of culture's influence on satisfying hunger is that
    a. only a portion of the potential food available in a society is defined as edible.
    b. people everywhere eat three meals a day.
    c. fast food appeals to people everywhere.
    d. if people are hungry enough, they will eat just about anything.

11. In English, the word *corn* denotes
    a. food.
    b. a gift.
    c. an important vegetable.
    d. the staple grain of a country.

12. ___________ are internal bodily sensations that we experience in relationships with other people.
    a. Social emotions
    b. Feeling rules
    c. Emotional states
    d. Expressive norms

13. Sociologist Choong Soon Kim found that _________ percent of Korean husbands and wives kissed each other upon being reunited after 30 years of separation.
    a. 0
    b. 10
    c. 30
    d. 50

14. Microwave ovens have
    a. caused the breakdown of the family.
    b. made it easier for people to eat at a set time with their families.
    c. eliminated some of the incentives for families to eat together.
    d. tied people to a tighter dining schedule.

15. Hangul is an important factor in explaining the ______ percent literacy rate in South Korea.
    a. approximately 85
    b. nearly 100
    c. 20
    d. 60

16. People of one society borrow ideas, materials, or inventions from another society if
    a. the items are exotic.
    b. the people admire the country from which the items are borrowed.
    c. the people perceive the items as useful.
    d. the items can be acquired at a reasonable cost.

17. The concept of selective borrowing has the following important implication for anyone trying to sell products to people who live in foreign markets:
    a. There is no need to customize products because people tend to accept foreign products.
    b. Marketers must convince people to purchase foreign products.
    c. People of one society borrow ideas indiscriminately from another society.
    d. The producer who doesn't tailor is the producer who will fail.

18. Which one of the following statements does not reflect a type of ethnocentrism?
    a. A foreign culture is perceived as the standard for judging the worth of a home culture.
    b. Outsiders deem a culture so offensive that they believe it must be destroyed.

c. The members of one culture believe so deeply in their ways that they have no concepts for thinking about other cultures.
d. A cultural practice is considered in light of its own cultural context.

19. _________ is the strain that people from one culture experience when they must reorient themselves to the ways of a new culture.
a. Culture shock
b. Ethnocentrism
c. Diffusion
d. Reverse ethnocentrism

20. The most extreme and most destructive form of ethnocentrism is
a. reverse ethnocentrism.
b. defining foreign ways as peculiar.
c. cultural genocide.
d. self-determination.

21. For much of the twentieth century, the Bureau of Indian Affairs used formal education as a tool
a. to wipe out Native American cultures.
b. to preserve Native American languages.
c. to teach Native Americans about their roots.
d. to bridge the gap between Native American youth and their families.

22. An individual who adopts cultural relativism aims to ________ a cultural practice.
a. understand
b. condone
c. discredit
d. judge

23. Which of the following is the best example of a distinct subculture?
a. Native Americans
b. Old Order Amish
c. South Koreans
d. Hispanics

24. Subcultures are ________ when their members do not interact with anyone outside their subculture to shop for food, attend schools, receive medical care, or find companionship because their needs are provided by their subculture.
    a. countercultures
    b. culturally complete
    c. subversive
    d. institutionally complete

Use one from the following set of responses for questions 25-27 to decide whether each statement is most likely to be made by
    a. a functionalist.
    b. a conflict theorist.
    c. a symbolic interactionist.

25. Culture represents the solutions people have worked out over time to meet environmental challenges.

26. Culture is an elaborate and complex system of meanings on which people have come to agree.

27. Owners of the means of production impose culture on other, less powerful groups.

28. According to the textbook, the role of culture in our lives is comparable to the role of __________. We rarely think about either, but we would be thrown off balance if either were to disappear.
    a. food
    b. gravity
    c. water
    d. education

29. According to anthropologist Edward T. Hall, Americans who are sent overseas
    a. are respected, even if they are not liked.
    b. usually are trained to speak the language and know the culture.
    c. usually are blind to the fact that what passes as ordinary, acceptable American behavior is not interpreted as such by foreigners.
    d. are interested in understanding the norms that govern communication in the foreign society.

30. Which one of the following reasons are sociologists least likely to put forth to explain why they are interested in culture?
    a. Knowledge of other cultures will help us to sell products to foreign consumers.
    b. Knowledge of other cultures teaches us that human behavior is constructed.
    c. Knowledge of other cultures offers us insights into the workings of American society.
    d. Knowledge of other cultures teaches us that behavior is not determined and that people need not be prisoners of their cultures.

True/False Questions

T F 1. Societies can be distinguished from each other on the basis of which values are present in one society and not in the other.

T F 2. If an English speaker wanted to convey to a Korean-speaking person that he or she adopted a child, that person would have known only the Korean word for adoption.

T F 3. What is considered funny in one culture is usually considered not funny in another.

T F 4. Much of Korean identity is tied up with being "not Japanese."

T F 5. Folkways are more important norms than mores.

T F 6. Part of the reason why Koreans and Americans use refrigerators and paper differently is connected with the amount of natural resources in each country.

T F 7. Our genetic endowment gives us our human, physical traits, and a culture.

T F 8. The opportunity to "borrow" from another culture occurs whenever two persons from different cultures make contact.

T F 9. On close analysis, many of the West's greatest achievements--modern agriculture, the printing press, the steam engine--are not achievements at all, but simple borrowings.

T F 10. The Japanese educational system is superior to the American system in almost everyway.

T F 11. Sociologists generally agree about how culture affects people's lives.

T F 12. The call for cross-cultural awareness was first made in the late 1980s and early 1990s.

Continuing Education

Leisure

For a change of music, listen to

Byon Kyu Man: Traditional and Contemporary Music from Korea. Adda.
P'Ansori: Korea's Epic: Vocal Art and Instrumental Music. Elektra/Nonesuch.

The next time you rent a movie, consider

Iron and Silk, directed by Shirley Sum (93 minutes). This film is based on the experiences of an American who goes to the People's Republic of China to teach English. Viewers experience Chinese culture through the eyes of the American director and at the same time confront many common American misconceptions about that culture (e.g., ideas about the martial arts). As you view the film, think about which of the cultural concepts considered in Chapter 4 apply to the events in this film.

Travel

If you're in St. Louis, visit the Dog Museum (1721 S. Mason Road; 314-821-3647). The items and displays in this museum show the various ways in which dogs have been depicted throughout history, and are good examples of culture's effect on the treatment of animals.

If you're in Portland, Oregon, visit the American Advertising Museum (9 NW Second Avenue at Couch Street; 503-226-0000). The museum contains more than 200,000 print advertisements, 14,000 slides of billboards and a library of 4,000 advertising books. Because the museum chronicles advertising in America over more than a century, you can learn much about the changes and the continuities in American values from viewing the collection.

General and Readable Information

Korea Journal (Korean National Commission for UNESCO, CPO Box 64, Seoul, 100-600 Korea) is a quarterly publication in English. Each issue contains editorials, letters, feature articles, poetry, fiction, and book reviews about all aspects of Korean culture.

Country Profile
South Korea

Official Name: Republic of Korea
Population (1994 est.): 44.5 million
Land Area: 38,120 square miles
Population Density: 1,166 people per square mile
Birth Rate per 1,000: 16
Death Rate per 1,000: 6
Rate of Natural Increase: 1.0%
Average Number of Children Born to a Woman During Her Lifetime: 1.6
Infant Mortality Rate: 15 deaths per 1,000 live births
Life Expectancy at Birth: 71
Gross National Product per Capita: $6,790
Capital: Seoul

"Korea was first populated by a Tungusic branch of the Ural-Altaic family, which migrated to the peninsula from the northwestern regions of Asia. Some also settled parts of northeast China (Manchuria); Koreans and Manchurians still show physical similarities-- in their height, for example. Koreans are racially and linguistically homogeneous, with no sizable indigenous minorities except Chinese (50,000).

South Korea's major population centers are in the northwest area of Seoul-Inchon and in the fertile southern plain. The mountainous central and eastern areas are sparsely inhabited. Between 1925 and 1940, the Japanese colonial administration in Korea concentrated its industrial development efforts in the comparatively underpopulated and resource-rich north, resulting in a considerable migration of people to the north from the southern agrarian provinces. This trend was reversed after World War II, when more than 2 million Koreans moved from the north to the south after the peninsula was divided into U.S. and Soviet military zones of administration. This southward migration continued after the Republic of Korea was established in 1948 and during the Korean war (1950-1953). About 10 percent of the people in the Republic of Korea are of northern origin. With 43 million people, South Korea has one of the world's highest population densities--much higher, for example, than India or Japan. While the territorially larger North Korea has only about 20 million people. Ethnic Koreans now residing in other countries live mostly in China (2.6 million), the United States (1.2 million), Japan (700,000), and the former Soviet Union (500,000)."

Sources: Haub and Yanagishita (1994); U.S. Department of State (1991).

Chapter References

Davis, Marilyn P. 1990. *Mexican Voices/American Dreams: An Oral History of Mexican Immigration to the United States*. New York: Henry Holt.

Haub, Carl and Machiko Yanagishita. 1994. *1994 World Population Data Sheet*. Washington, DC: Population Reference Bureau.

Hogan, Mary Ann. 1994. "The Joy of Crying." *Los Angeles Times* (February 1):E1+.

Japan Information Center. 1988. *What I Want to Know about Japan*. New York: Consulate General of Japan.

NPR. 1992. "Morning Edition." (0ctober 8):36.

Perry, Charles. 1994. "Peanut Butter: A Stirring Tale." *Los Angeles Times* (March 10):H9+.

U.S. Department of State. 1991. "South Korea." *Background Notes* (#7782). Washington, DC: U.S. Government Printing Office.

Answers

Concept Application

1. Reverse ethnocentrism; Material culture; Diffusion
2. Feeling rules
3. Ethnocentrism
4. Diffusion
5. Norms

Multiple-Choice

1.b 2.a 3.a 4.d 5.d 6.a 7.d 8.c 9.c 10.a 11.d 12.a 13.a 14.c 15.b 16.c 17.d 18.d 19.a 20.c 21.a 22.a 23.b 24.d 25.a 26.c 27.b 28.b 29.c 30.a

True/False

1.F 2.F 3.F 4.T 5.F 6.T 7.F 8.T 9.T 10.F 11.F 12.F

Chapter 5
Socialization

Study Questions

1. What is socialization?

2. What are the basic dynamics underlying the century-long struggle between Arabs and Jews? With regard to socialization, what questions would sociologists find interesting to ask about this conflict?

3. What is the relationship between nature and nurture?

4. How do extreme cases of isolation underscore the importance of socialization? Use one of the following cases to illustrate: (a) Anna and Isabelle; (b) children orphaned as result of the Holocaust; (c) Spitz's study of orphanages for children of prison mothers; and (d) the elderly in nursing homes.

5. On the basis of Anna's and Isabelle's case histories, what conclusions did Kingsley Davis reach about the effects of prolonged isolation? What factor did Davis overlook in drawing his conclusions?

6. What is the social importance of memory?

7. What are primary groups? How are they important agents of socialization?

8. What characteristics make a military unit a primary group?

9. According to sociologists Donald E. Miller and Lorna Touryan Miller, how is memory transmitted across generations?

10. What are ingroups and outgroups? What is the sociological significance of ingroups and outgroups?

11. According to George Herbert Mead, what are the mechanisms that allow an individual to interact with others?

12. Distinguish between the "I" and the "me." How does the "me" develop?

13. How do children come to learn to take the role of others?

14. According to Charles Horton Cooley's "looking-glass self" theory, how does a sense of self develop?

15. What central concept underlies Piaget's theory of cognitive development?

16. What is resocialization? What are the types of resocialization?

17. What mechanisms do total institutions use to resocialize inmates?

18. Under what conditions are people least likely to resist resocialization?

19. Expand on the following idea: "Human genetic and social makeup contains considerable potential for change."

Concept Application

Below are five scenarios and the sources from which they were drawn. Decide which concept or concepts covered in Chapter 5 are represented best by each scenario, and explain why. The following concepts are considered:

Active adaptation
Collective memory
Engram
Games
Generalized other
Group
Ingroup
Internalization
Looking-glass self
Nature
Nurture
Outgroup
Play
Primary groups
Reflexive thinking
Resocialization
Role-taking
Significant others
Significant symbols
Symbolic gestures
Total institutions

Scenario 1

"In 1910, two French surgeons wrote about their successful operation on an 8-year-old boy who had been blind since birth because of cataracts. When the boy's eyes were healed, they removed the bandages, eager to discover how well the child could see. Waving a hand in front of the boy's physically perfect eyes, they asked him what he saw. He replied weakly, 'I don't know.' 'Don't you see it moving?' they asked. 'I don't know' was his only reply. The boy's eyes were clearly not following the slowly moving hand. What he saw was only a varying brightness in front of him. He was then allowed to touch the hand as it began to move; he cried out in a voice of triumph: 'It's moving!' He could feel it move, and even, as he said, 'hear it move,' but he still needed laboriously to learn to *see* it move" (Zajonc 1993, p. 22).

Scenario 2

"I see Israeli Arabs almost everywhere I go, and some Palestinians. I search their faces, wanting to know something about their lives. But for me and for many tourists, contact here is more infrequent and impersonal, and tinged by a certain weariness that I recognize and do not like. I feel like a Peeping Tom along the streets of Acre and Nazareth. At the Western Wall, Muslims in dress clothes pass on their way to and from mosques, and I wonder what they think as they pass us. Throughout the country, I am warned about where we should and should not go near the end of Ramadan, Islam's holy month" (Torgovnick 1993, p. xx31).

Scenario 3

"Let me say to you, the Palestinians, we are destined to live together on the same soil in the same land.

We, the soldiers who have returned from battles stained with blood; we who have seen our relatives and friends killed before our eyes; we who have attended their funerals and cannot look into the eyes of their parents; we who have come from a land where parents bury their children; we who have fought against you, the Palestinians" (Rabin 1993, p. A7).

Scenario 4

"Genetic endowments may set limits for the height or intelligence that individuals can attain but their actual height or intelligence also depends upon how they are raised. The increasing height of the American population over the past several generations reflects the change in nutritional conditions and probably the diminution in childhood illnesses more than a genetic selection" (Lidz 1976, p. 40).

Scenario 5

"There were three tiers of close family. At the top were remote grownups--my father and mother and even more distant Uncle Simon and Aunt Sarah. Next in age to my parents, uncle and aunt came a 'middle' group of my older sister Min, my half-brother Adolf, and two cousins who were of an age ten or fifteen years older than we were. 'We' were the four of us. We were the center of my world: until I was six or seven, my sister Alice, not quite two years older than I, hot-tempered and affectionate, and my cousins Marvin and Julia" (Bruner 1983, p. 10.).

Applied Research

Below is a selective list of various locations around the world in which there are "ethnic conflicts" (Binder and Crossette 1993, 12y). Find a newspaper article, a passage in a book, a radio segment, a TV report, a movie, or other information on how some aspect of socialization plays a role in perpetuating this type of conflict.

Location

Bosnia and Herzegovina
Croatia
Spain
United Kingdom
Romania
Moldova
Azerbaijan
Iraq
Egypt

Sudan
Mauritania
Senegal
Liberia
Togo
Nigeria
Uganda
Rwanda
Burundi
India
Sri Lanka
Myanmar
Cambodia
Peru

Source: (Binder and Crossette 1993, p. 12Y).

Practice Test

Multiple-Choice Questions

1. Socialization is a process that
   a. begins at around two years of age.
   b. lasts about 25 years.
   c. begins at birth.
   d. occurs off and on throughout the life cycle.

2. The Palestinian-Israeli conflict has lasted
   a. a century.
   b. a decade.
   c. a half-century.
   d. 10,000 years.

3. The holocaust claimed the lives of more than 6 million Jews. That figure represents approximately _________ of European Jews.
   a. one-half

b. 90 percent
c. one-third
d. 10 percent

4. With the exception of __________, Israelis and Palestinians have little in common.
   a. an economic relationship
   b. religious ties
   c. military service
   d. shared neighborhoods

5. One important factor that led to the 1987 intifadah movement was
   a. a severe and prolonged drought.
   b. Israeli military occupation, which began in 1985.
   c. the Six-Day War, in which Israel defeated six Arab armies.
   d. massive seizures by the Israeli government of Palestinian agricultural and grazing land.

6. _______ is the term for human genetic makeup or biological inheritance.
   a. Nature
   b. Nurture
   c. Internalization
   d. Socialization

7. Perhaps the most outstanding feature of the human brain is its
   a. size.
   b. shape.
   c. flexibility.
   d. color.

8. In "Two Cases of Extreme Isolation," sociologist Kingsley Davis speculated on why Anna did not achieve a level of thought and behavior normal for her age, as did Isabelle. Which one of the following explanations for this difference did Davis overlook?
   a. Anna may have inherited a physical and mental constitution that was less hardy than Isabelle's.
   b. Anna did not receive the intensive and systematic therapy that Isabelle received.

c. Anna died at age 10; consequently we will never know whether she could have achieved a state of normal development.
d. Unlike Isabelle, Anna was shut off from everyone, including her mother.

9. A bond of mutual expectation is established between caregiver and baby when
   a. a caregiver knows the baby well enough to understand its needs and feelings.
   b. the baby is put on a strict feeding and sleeping schedule.
   c. the baby learns to talk to the caregiver.
   d. the caregiver can leave the baby alone in a room without the baby's crying.

10. _______ is the concept used to describe the experiences shared and recalled by significant numbers of people.
    a. Engrams
    b. Collective memory
    c. National memory
    d. Cohort time

11. From a sociological viewpoint, memory of past experiences is important because it
    a. shapes individuals' unique character.
    b. allows individuals to participate in society and shape their environment.
    c. allows people to store experiences unique to their lives.
    d. sheds light on the biological quality of memory.

12. _______ are major socializing agents, especially in the early years.
    a. Primary groups
    b. Ingroups
    c. Secondary groups
    d. Cohorts

13. Which one of the following is not a primary group?
    a. Military unit
    b. Family
    c. Basketball team
    d. Girl Scouts of America

14. Sociologists Amith Ben-David and Yoah Lavee's study, in which they describe how Israeli family members interacted with one another during the SCUD missile attack, shows that the family
    a. can serve to buffer its members against the effects of negative circumstances, or can exacerbate those effects.
    b. is a supportive and positive influence during a crisis.
    c. increases the stress of a crisis.
    d. becomes divided and tense during a crisis.

15. Those groups with which people identify and to which they feel closely attached, particularly when that attachment is founded on hatred of another group, are
    a. essential groups.
    b. respected groups.
    c. outgroups.
    d. ingroups.

16. Jews from at least 80 different countries have settled in Israel. Which of the following is not one of the factors that unify the culturally diverse Israeli population?
    a. Shared language
    b. A shared desire for a homeland free of persecution
    c. The ongoing conflict with the Palestinians
    d. The universally shared attitude that there be no compromise with the Palestinians

17. In general, many of the clashes between Palestinians and Jews are over
    a. language.
    b. symbols.
    c. religion.
    d. status differences.

18. When Israeli and Palestinian children dream about meeting one another, the other appears in their dreams as
    a. a friend.
    b. a classmate.
    c. a terrorist or an oppressor.
    d. a relative.

19. The emergence of self depends on our physiological capacity for reflexive thinking, which is
    a. the ability to step outside the self and observe and evaluate it from another's viewpoint.
    b. the environment or interaction experiences beginning at birth.
    c. the capacity to learn a spoken language.
    d. the ability to recall memories stored in engrams.

20. Two important kinds of significant symbols are
    a. symbolic gestures and body language.
    b. words and definitions.
    c. language and symbolic gestures.
    d. tone of voice and facial expression.

21. According to George Herbert Mead, the "me" is the part of the self that
    a. is spontaneous and creative.
    b. acts in unconventional ways.
    c. develops through imitation, play, and games.
    d. is capable of rejecting expectations.

22. According to George Herbert Mead, children acquire a sense of self when they are able to
    a. remember.
    b. role-take.
    c. imitate others.
    d. repeat important phrases.

23. Palestinian children who pretend to be Israeli soldiers arresting and beating stone throwers are in the
    a. preparatory stage.
    b. play stage.
    c. game stage.

24. Which one of the following statements is false with regard to Charles Horton Cooley's conception of the looking-glass self?
    a. We see ourselves reflected in others' reactions to our appearance and behaviors.
    b. We acquire a sense of self by being sensitive to the appraisals of us that we perceive others to have.

c. The imagining or interpreting of others' reactions is critical to self-awareness.
d. People respond to others' actual reaction to them.

25. Piaget believed that the ability to learn and reason is rooted in
a. active adaptation.
b. play.
c. imitation.
d. games.

26. Which one of the following statements is a characteristic of Piaget's model of cognitive development?
a. Cognitive development involves three broad stages.
b. A child can proceed from one stage to another, even if reasoning challenges of an earlier stage are not mastered.
c. A more sophisticated level of cognitive understanding will not show itself until the brain is ready.
d. The theme of ingroup-outgroup runs through all stages.

27. People who choose to participate in a process or program designed to remake them undergo ________ resocialization.
a. systematic, voluntary
b. involuntary
c. informal, systematic
d. voluntary, informal

28. Mental hospitals, concentration camps, boarding schools, and monasteries are
a. organizations.
b. secondary groups.
c. primary groups.
d. total institutions.

29. What we know about socialization and resocialization processes suggests that
a. events unfold in a predictable fashion.
b. human genetic and social makeup contains considerable potential for change.
c. no generation can do much to change the problems it inherits from previous generations.
d. people cannot be resocialized to abandon one way of thinking and behaving for another.

30. From a conflict perspective, the kinds of clothes students wear
    a. send a message about where they fit in, where they want to fit in, and whether they fit in at all.
    b. perpetuate the "us" versus "them" mentality.
    c. encourage them to strive to obtain success in life.

True/False Questions

T F 1. Memories are stored in engrams (physical traces in the brain) much as films are stored on videocassettes.

T F 2. Palestinians and Israelis have relied largely on resocialization measures that attempt to force the other side to change.

T F 3. Biology is destiny.

T F 4. Nature is more important than nurture.

T F 5. In their study of 100 survivors of the Armenian genocide, Miller and Miller found that grandparents are primary carriers and transmitters of collective memory.

T F 6. George Herbert Mead was very specific about how the "I" develops.

T F 7. Most "newcomers" to a society become carbon copies of their teachers.

T F 8. Meaningful social contact and stimulation from others are important at any age.

T F 9. The consequences of extreme isolation for the institutionalized elderly are strikingly similar to those described in the late 1940s by Spitz, who studied orphans.

Continuing Education

Leisure

For a change in music, listen to

Jewish-Yiddish Music
Beautiful Israel: Golden Folk Songs (Gevatron and the Effi Netzer Singers). Fiesta.
Desert Wind (Ofra Haza). Sire.
Shalom: Folklore and New Songs of Israel. Calig.
Partisans of Vilna: The Songs of World War II (Jewish Resistance). Flying Fish.

Palestinian Music
The Death of a Prophet (Maut Nabe).
Bide My Time (Subreen).
The Land of Amer's Sun (Mal Ben-Amer).

The next time you rent a movie, consider

The Emerald Forest, directed by John Boorman (113 minutes). This film is based on a true story of a boy kidnapped at the age of 7 or 8 by a tribe living in the Amazon. The boy's father searches 10 years before he finds his son. By that time the son has become completely comfortable with the forest and the culture of the tribe he has been living with. The son, now a young man, decides to not return "home" with his father. As you view the film, think about the tremendous influence of "nurture" on the development of self. *The Emerald Forest* not only illustrates the dynamic forces of socialization on the self; it also shows how the destruction of the Amazon is affecting the culture of the Amazon tribes and is disrupting the relationships among various tribal groups living in that area.

Travel

If you're in Chicago, visit the Oriental Institute Museum (1155 E. 58th Street; 312-702-9520). It displays examples of architecture, art, religion, and daily life in the ancient Near East, including Palestine.

If you're in New York City (Upper East Side), visit the Jewish Museum (1109 Fifth Avenue at E. 92nd Street; 212-860-1888). This museum is the largest of its kind in the world.

If you're in Atlanta, visit the Toy Museum of Atlanta (2800 Peachtree Road, NE; 404-266-8697). Approximately 100,000 toys from the 19th century are housed here. The exhibits help us to see the importance of toys (their nature and purpose) in child socialization.

If you're in Washington, DC, visit the United States Holocaust Memorial Museum (100 Raoul Wallenburg Place, SW; 202-488-0400), an architectural memorial and a museum devoted to preserving the memory of the Holocaust.

General and Readable Information

Middle East Insight (1200 18th Street NW, Suite 305, Washington, DC 20036). According to the magazine's advertisement, "Each issue contains lively, stimulating, and timely analyses and opinions from across the spectrum of Middle East perspectives and problems by the best minds in the field. *Middle East Insight* is unique in its mission to promote understanding by providing an impartial forum for the serious presentation of diverse views."

Country Profile
Israel

Official Name: State of Israel
Population (1994 est.): 5.4 million
Land Area: 7,850 square miles
Population Density: 692 people per square mile
Births per 1,000: 21
Deaths per 1,000: 6
Rate of Natural Increase: 1.5
Average Number of Children Born to a Woman During Her Lifetime: 2.8
Infant Mortality Rate: 8.1 deaths per 1,000 live births

Life Expectancy at Birth: 76 years
Gross National Product per Capita: $13,230
Capital: Jerusalem

Country Profile
Gaza

Official Name: NA
Population (1994 est.):
Land Area: 2,123 square miles
Population Density: NA
Birth Rate per 1,000: 56
Death Rate per 1,000: 6
Rate of Natural Increase: 5.0%
Average Number of Children Born to a Woman During Her Lifetime: 7.7
Infant Mortality Rate: 43 deaths per 1,000 live births
Life Expectancy at Birth: 66 years
Gross National Product per Capita: $850
Capital: None

Country Profile
West Bank

Official Name: NA
Population (1994 est.):
Land Area: 140 square miles
Population Density: NA
Birth Rate per 1,000: 46
Death Rate per 1,000: 7
Rate of Natural Increase: 4.0%
Average Number of Children Born to a Woman During Her Lifetime: 5.7
Infant Mortality Rate: 40 deaths per 1,000 live births

Life Expectancy at Birth: 69 years
Gross National Product per Capita: NA
Capital: none

Of the approximately 5.2 million Israelis in 1992, about 4.2 million were Jewish. While the non-Jewish minority grows at an average rate of 4% per year, the Jewish population has increased by 10% over the last three years as a result of massive immigration to Israel, primarily from the republics of the former Soviet Union. In the past three years, nearly 400,000 such immigrants arrived in Israel, making this the largest wave of immigration since independence. In addition, almost 20,000 members of the Ethiopian Jewish community have immigrated to Israel, 14,000 of them during the dramatic May 1991 Operation Solomon airlift.

The three broad Jewish groupings are: the Ashkenazim, or Jews who came to Israel mainly from Europe, North and South America, South Africa, and Australia; the Sephardim, who trace their origin to Spain, Portugal, and North Africa; and Eastern or Oriental Jews, who descend from ancient communities in Islamic lands. Of the non-Jewish population, about 77% are Muslims, 13% are Christian, and about 10% are Druze and others.

Sources: Haub and Yanagishita (1994); U.S. Department of State (1994).

Chapter References

Binder, David and Barbara Crossette. 1993. "As Ethnic Wars Multiply, U.S. Strives for a Policy." *The New York Times* (February 7):Y1+.

Bruner, Jerome. 1983. *In Search of Mind: Essays in Autobiography*. New York: Harper & Row.

Haub, Carl and Machiko Yanagishita. 1994. *1994 World Population Data Sheet*. Washington, DC: Population Reference Bureau.

Lidz, Theodore. 1976. *The Person: His and Her Development Throughout the Life Cycle*. New York: Basic Books.

Rabin, Yitzhak. 1993. "Making a New Middle East. 'Shalom, Salaam, Peace.' Views of Three Leaders." *Los Angeles Times* (September 14):A7.

Torgovnick, Marianna. 1993. "Carrying Emotional Baggage to Israel." *The New York Times* (July 25):xx31.

U.S. Department of State. 1994. "Israel." *Background Notes* (#7752). Washington, DC: U.S. Government Printing Office.

Zajonc, Arthur. 1993. "Seeing the Light." *Los Angeles Times Magazine* (July 25):22-25.

Answers

Concept Application

1. Nature; Nurture; Resocialization
2. Outgroup
3. Collective memory
4. Nature and Nurture
5. Primary groups; Significant others

Multiple Choice

1.c 2.a 3.c 4.a 5.d 6.a 7.c 8.d 9.a 10.b 11.b 12.a 13.d 14.a 15.d 16.c 17.b 18.c 19.a 20.c 21.c 22.b 23.b 24.d 25.a 26.d 27.a 28.d 29.b 30.b

True/False

1.F 2.T 3.F 4.F 5.T 6.F 7.F 8.T 9.T

Chapter 6
Social Interaction and the Social Construction of Reality

Study Questions

1. What is social interaction? How do sociologists approach social interaction?

2. How is Zaire related to issues of social interaction and the social construction of reality?

3. How did Durkheim define the division of labor? How is the division of labor related to global interdependence?

4. Distinguish between organic and mechanical solidarity.

5. What kinds of disruptive events break down the abilities of individuals to connect with one another in meaningful ways through their labor?

6. Use the case of Zaire to give an example of each of the five disruptive events.(An example may be used more than once.)

7. How is Dr. Rask's biography connected to these disruptive events?

8. How is the activation and spread of HIV connected to the unprecedented mixing of people from all over the world?

9. Who is responsible for triggering and transmitting HIV? Explain your answer.

10. How can people interact smoothly with people they know nothing about? Explain.

11. What is a social status? How is it related to social structure?

12. What is the difference between achieved and ascribed statuses? Is the difference clear-cut? Why or why not?

13. How are the concepts of status, role set, rights, and obligations related?

14. Does the idea of role imply totally predictable behavior? Explain.

15. What are the broad differences between practitioner-patient roles in Africa and in the United States?

16. How was the content of Dr. Rask's interactions with physicians and friends shaped by cultural expectations about the roles of physician and of patient?

17. What is impression management? What interaction dilemmas are associated with impression management?

18. How does the concept of impression management help us to understand the dilemmas health leaders face in the United States and Zaire in convincing people to use condoms?

19. What is the difference between backstage and frontstage? Use these concepts to analyze blood banks and their handling of HIV.

20. What sociological concepts would you draw upon to analyze the content of interaction?

21. People usually attribute cause to either dispositional traits or situational factors. What is the difference between dispositional traits and structural factors? Give an example of each.

22. What problems are associated with using dispositional traits to explain the cause of AIDS and to diagnose AIDS cases?

23. What must take place before we can truly understand the cause of HIV and AIDS?

24. What are some of the strategies people use to construct reality?

25. Explain the following statement: "Television uses an image-oriented format." What are the problems associated with such a format?

26. Describe the ways young black males manage the setting, their dress, their words, and/or their gestures to make a favorable impression on police officers.

Concept Application

Below are five scenarios and the sources from which they were drawn. Decide which concept or concepts covered in Chapter 6 are represented best by each scenario, and explain why. The following concepts are considered:

Achieved status
Ascribed status
Backstage
Content (of interaction)
Context (of interaction)
Dispositional traits
Division of labor
Dramaturgical model
Frontstage
Impression management
Mechanical solidarity
Obligations

Organic solidarity
Rights
Role
Role conflict
Role set
Role strain
Scapegoat
Sick role
Situational factors
Social interactions
Social status
Social structure
Solidarity
Stigmas

Scenario 1

"I use a wheelchair because I was paralyzed by polio 40 years ago. One of my first trips out of the hospital back then was to a supermarket. I remember I was rolling down an aisle when a kid saw me. He stopped dead in his tracks and pointed. 'Mommy,' he said in a loud voice--a very loud voice, 'Mommy, look at the broken man'" (Gallagher 1992, p. 17).

Scenario 2

"Ten minutes after William Andrews succumbed to the poisonous concoction injected into his arm, Dr. Robert Jones performed a task from which, he said, he would never quite recover: He entered the chamber of death, checked the condemned man's vital signs and confirmed that he was, in fact, dead.

The medical director for the Utah State Prison system did not witness the July, 1992, execution. But his limited role so troubled him that he decided never again to have anything to do with a state-ordered killing.

'It was much more stressful, much more disconcerting than I thought it would be,' Jones says. 'I literally slept for a whole day afterward, and I thought, 'That's an experience in life that you don't want to have to go through again....Physicians usually try to preserve life, not end it.'

As a prison doctor, Jones sits at the uncomfortable intersection of medicine and criminal justice. His dilemma highlights an ethical debate that is raging in the medical community: Should doctors, who take the Hippocratic oath not to harm their patients, take part in carrying out the death penalty? When state laws and regulations require physicians to be present at executions--as in California, where doctors watch the heart monitor that charts the prisoner's final moments in the gas chamber--should the physician comply?" (Stolberg 1994, p. E1).

Scenario 3

"It was early on a cool spring day, but in the dark subway corridors beneath Flatbush, there was no way to know it. Officer Michael K., a black transit police officer, was in full plainclothes outfit: sneakers, jeans, sweatshirt. His knee was pressed to the chest of a suspect whom he had wrestled to the ground, and his gun was pushed against the man's temple.

Then, out of the corner of his eye he saw something move. A white man, in combat stance, was facing him, pointing a gun directly at him. Officer K. screamed out, 'Joe! Stop! It's me!' Somehow, in that half second, the man's knees gave out and he lowered his gun.

Officer K. remembered one more thing, vividly: the white man, a fellow officer, stumbled toward him. 'Oh my God, Mike,' he cried out, his jaw shaking. 'I didn't know. All I saw was a black guy with a gun.'

By now, Officer K.'s experience, which he related on Thursday on the condition that his full name not be used, is a tale with familiar trappings. Last Tuesday in Brooklyn, a black transit police officer making an arrest was shot and seriously wounded by a white colleague who mistook him for a mugger" (Wolff 1992, p.Y1).

Scenario 4

"The average length of hospitalization in Japan is by far the longest in the world....One of the major factors...is a basic "pampering" attitude toward the sick....The Japanese think of many more conditions as 'illnesses' than biomedicine recognizes as 'diseases.'...

Part and parcel of the Japanese attitude toward illness is the emphasis on *ansei* (peace and quiet, or bed rest) as the major treatment for virtually any illness, from a minor cold to a major disease....Lengthy hospitalization, then, may be seen as the official sanctioning of *ansei*, the most cherished treatment method of popular Japanese medicine. Long hospitalization periods thus affirm the legitimacy of, and provide the institutional support for, sickness....

For both men and women, there is an implicit and sometimes explicit expectation on the part of the patient, approved by family members and doctors, that hospitalization is a form of 'vacation,' a reward for hard work. This attitude may be linked to the limited use of paid vacation time by Japanese employees. Illness legitimizes the 'vacation' Japanese workers otherwise feel pressured not to utilize. Needless to say, the connection between these two phenomena is not always consciously recognized" (Fox 1989, p. 20).

Scenario 5

"Gergen remembers the selling of the Nixon image as a central fact of life in his first White House. 'The great cynicism had already begun when Nixon took office,' he says. 'It came with Vietnam when a generation of reporters concluded that their Government was lying to them. Nixon arrived on the back end of that, and he came carrying this great personal animosity, a lack of trust in the press.'

And Nixon understood that unless you mastered television, your Presidency would be ripped apart. And so he developed, the Nixon White House developed, a whole series of ideas about how one talks to the press and communicates through the press. A sort for rules of the road for how a White House acts'" (Kelly 1993, p. 67).

Applied Research

In this chapter we considered Emile Durkheim's theory of the division of labor. Recall that Durkheim hypothesized that societies become vulnerable as the division of labor becomes more complex and more specialized. He was particularly concerned with the kinds of events that break down peoples' abilities to connect with one another in meaningful ways through their labor. In Chapter 6 we showed how such disruptions were connected to the transmission of HIV. Apply Durkheim's theory to the case of Romania. For background information see Rothman and Rothman (1990; 1993).

Practice Test

Multiple-Choice Questions

1. When sociologists study social interaction, they seek to understand and explain
   a. biography and history.
   b. context and content.
   c. the personalities of the people involved.
   d. the interrelationship between genetic makeup and culture.

2. HIV existed as early as
   a. 1959.
   b. 1969.
   c. 1980.
   d. 1975.

3. Zaire is emphasized in Chapter 6 ("Social Interaction and the Social Construction of Reality") because
   a. HIV originated in Zaire.
   b. HIV "traveled" from Europe to Zaire.
   c. an unidentified blood sample frozen in 1959 and stored in a Zairian blood bank provides evidence that HIV existed before the 1980s.
   d. that country is located in Africa, the so-called "cradle of AIDS."

4. ___________ wrote *The Division of Labor in Society.*
   a. Karl Marx
   b. Max Weber
   c. Emile Durkheim
   d. C. Wright Mills

5. *The Division of Labor in Society* offers a framework for understanding all but which one of the following issues?
   a. Global interdependence
   b. Conditions that cause large-scale social upheaval
   c. Social conditions that leave people vulnerable
   d. The specific origin of HIV

6. Solidarity is a term used by Durkheim to refer to
   a. a characteristic of hunting and gathering societies.
   b. the ties that bind people to one another in a society.
   c. specialization of work tasks.
   d. mechanization.

7. The Mbuti pygmies are
   a. a hunting and gathering people who moved from the forest to the cities.
   b. a nation of people who share a forest-oriented value system.
   c. a hunting and gathering people with ties characterized by organic solidarity.
   d. a nation of people that has become extinct.

8. A society with a complex division of labor is characterized by
   a. mechanical solidarity.
   b. common conscious.
   c. similarity and interdependence.
   d. organic solidarity.

9. Since 1883, when Zaire was colonized, the country has experienced a number of events that have weakened its solidarity. Which one of the following is not one of these events?
   a. Industrial and commercial crises
   b. Occupations filled on the basis of ascribed characteristics
   c. Unemployment and irregular work
   d. A shortage of natural resources

10. When the Belgians pulled out of Zaire in 1960, they left the country with only 120 medical doctors for 33 million people. This outcome can be traced to which one of the following disruptions in the division of labor?
    a. Industrial and commercial crisis
    b. Workers' strikes
    c. Job specialization
    d. Forced division of labor

11. Dr. Rask came to Zaire when
    a. Mobutu invited professional people from Denmark and other countries to work in Zaire.
    b. she lost her license to practice medicine in the United States.
    c. the Belgian government colonized Zaire.
    d. she found out that there was an AIDS epidemic.

12. In Asian countries (Pattern III), most cases of HIV occur among
    a. homosexual males.
    b. IV drug users.
    c. those who received imported blood that was HIV-infected.
    d. heterosexuals.

13. Mad cows disease can be traced to
    a. field mice.
    b. a single change in a manufacturing process that changed the purity of cattle feed.
    c. mosquitoes that hitched a ride on tires shipped from Japan to England.
    d. an American scientist who buried a pot of virus-laden soil in a field.

14. A social status is
    a. a role.
    b. a rank.
    c. a behavior.
    d. a position in a social structure.

15. __________ are statuses that are deeply discrediting because they overshadow all other statuses that a person occupies.
    a. Role conflicts
    b. Ascribed statuses

c. Stigmas
d. Achieved statuses

16. The sociological significance of status and role is that both concepts
    a. alert us to the fact that there is always room for improvisation and personal style.
    b. suggest that it is possible for people to interact with others without knowing them.
    c. alert us to issues of individual personality and interpret action.
    d. focus our attention on status conflict and status strain.

17. Which one of the following traits distinguishes Western-trained physicians from traditional healers?
    a. Healers concentrate on finding a cure, not on the relief of symptoms.
    b. Healers rely on drugs, not on surgery, to cure a condition.
    c. Healers attach considerable importance to factors other than biology, such as social relationships.
    d. Social interaction is utilitarian rather than personal.

18. Western-trained physicians working in Zaire's urban hospitals
    a. are more successful than traditional healers.
    b. are less likely to succeed if they consider other models of sickness.
    c. are likely to achieve better health outcomes if the patient's relatives are kept on the sidelines.
    d. are most successful when they tolerate, respect, and consider other models of illness.

19. Which one of the following statements about impression management is false?
    a. People usually are not aware that they are engaged in impression management because they are simply behaving in ways they regard as natural.
    b. Impression management can be a constructive and normal feature of social interaction.
    c. The dark side of impression management emerges when people manipulate the audience in deliberately deceitful and hurtful ways.
    d. If people spoke and behaved as they pleased, relationships would become more open.

20. From a global perspective, programs designed for women to prevent HIV infection must consider
    a. the importance of motherhood for many women around the world.

b. that most men take primary responsibility for preventing pregnancy.
c. women are not at risk of contracting HIV.
d. that most women have easy access to birth control.

21. The frontstage is the area
    a. out of the audience's sight.
    b. where people take care to create and maintain expected images and behavior.
    c. where individuals can "let their hair down."
    d. that people take great care to conceal from the audience.

22. ________ is a theoretical approach that helps us to understand how we arrive at our everyday explanation of behavior.
    a. Durkheim's theory of the division of labor
    b. The dramaturgical model
    c. Role-taking
    d. Attribution theory

23. People usually attribute cause to either dispositional traits or situational factors. Dispositional traits include
    a. bad luck.
    b. environmental conditions.
    c. motivation level.
    d. forces outside an individual's control.

24. Throughout history, whenever medical people haved lacked knowledge or technology to control a disease, the people in the society have tended to
    a. take humane and responsible steps to combat the disease.
    b. focus on the biological cause of the disease.
    c. act as if a cure will never be found.
    d. hold some group responsible for causing it.

25. The spread of HIV infection in Zaire is connected to all but which one of the following modes of transmission?
    a. Mother-to-child
    b. Reuse of needles for medical purposes
    c. Polygynous marriages
    d. Heterosexual contact

26. Western and African popular theories of the origins of HIV transmission are alike in that they tend to emphasize
    a. situational factors.
    b. dispositional traits.
    c. a homosexual connection.
    d. backstage factors.

27. In the West, an HIV-positive man who has received a blood transfusion and who has had sexual relations with another man is placed in which transmission category?
    a. Blood recipient
    b. Homosexual
    c. Blood recipient and homosexual
    d. Other

28. __________ are/is the most dependable method for determining the number of HIV-infected persons.
    a. Estimates
    b. Questionnaires
    c. Participant observation
    d. Random sampling of blood

29. This is one important question that may provide important clues to understanding AIDS:
    a. Why do young people refuse to use condoms?
    b. Why do some people remain HIV-infected for years and yet have never developed AIDS, while other people develop AIDS shortly after exposure to HIV?
    c. What is it about homosexuals that causes them to engage in high-risk behaviors?
    d. What country is the cradle of AIDS?

30. The format of television news gives viewers
    a. the impression that the world is manageable.
    b. the impression that the events covered can be explained.
    c. only the most superficial facts about important events going on in the world.
    d. time during commercials to reflect on the news they have heard.

31. Young black men who try to control the setting by their dress, their words, and/or their gestures when in the presence of police officers are engaged in
    a. role-playing.

b. structural strain.
c. backstage behavior.
d. impression management.

True/False Questions

T F 1. The content of interaction is related to the larger historical circumstances that bring people together.

T F 2. Mark Twain wrote "King Leopold's Soliloquy," an essay critical of Belgian imperialism.

T F 3. Disease patterns historically are affected by changes in population density and transportation patterns.

T F 4. Traditional healers attach almost no importance to the physical aspects of diseases.

T F 5. We know that 50 percent of hemophiliacs were HIV-infected by Factor VIII treatments before the first case of AIDS appeared in that group.

T F 6. When evaluating the causes of their own behaviors people tend to favor situational factors.

T F 7. The breakdown in sexual restraints and control is unique to Africa.

T F 8. Contemporary sexual life in many African towns and cities is influenced by Western norms that support free sex.

T F 9. From a sociological viewpoint, it is important to pinpoint the site of the first case of AIDS.

T F 10. Achieved and ascribed statuses are clear-cut categories.

T F 11. Attribution theory rests on the assumption that people draw on historical, cultural, and biographical information before assigning cause.

T F 12. In the United States, AIDS is confined to male homosexuals and IV drug users.

T F 13. HIV originated in Zaire and spread to the United States via Europe, Haiti, or Cuba.

Continuing Education

Leisure

For a change in music, listen to

Zaire: Musiques Urbaines a Kinshasa (various orchestras). Ocora.

The next time you rent a movie, consider

Being There, a film directed by Hal Ashby and based on a novel of the same title by Jerzy Kosinski. The central character is a gardener whose only knowledge of the world beyond the garden he tends comes from television. The film's sociological value lies in its lessons about interaction dynamics. Once people think they know the status of the main character (Peter Sellers), they proceed to interact on the basis of that status, making everything Sellers says or does consistent with the expectations associated with status. The movie also addresses issues concerning the social construction of reality, especially with regard to American politics.

Travel

If you're in Chicago, visit the International Museum of Surgical Sciences (1524 N. Lake Shore Drive; 312-642-3555). Surgical instruments from prehistoric times to the present are on display.

If you're in Washington, DC, visit the National Museum of African Art (950 Independence Avenue SW; 202-357-4600). This museum opened in 1987 and has approximately 6,000 items of African art from everywhere on the African continent.

General and Readable Information

Africa News (P.O. Box 6884, Syracuse, NY 13217-7917) is a bi-weekly publication that covers news about every country on the African continent. African News is a service agency supplying news and feature material to broadcast and print media.

BBC Focus on Africa (Bush House P.O.Box-Strand, London WC2B 4PH, U.K.) is a quarterly publication produced by the BBC African Service and its listeners. Each issue includes news features and updates chronicling important events whcih have occurred since the last issue. It also publishes short stories, poetry, letters to the editor, and articles of general interest.

Country Profile
Zaire

Official Name: Republic of Zaire
Population (1994 est.): 42.5 million
Land Area: 875,520 square mile
Population Density: 49 people per square miles
Birth Rate per 1,000: 48
Death Rate per 1,000: 15
Rate of Natural Increase: 3.3
Average Number of Children Born to a Woman During Her Lifetime: 6.7
Infant Mortality Rate: 93 deaths per 1,000 live births
Life Expectancy at Birth: 52 years
Gross National Product per Capita: NA
Capital: Kinshasa

The population of Zaire was estimated at 32 million in 1987. About 2,500 U.S. citizens live there. As many as 250 ethnic groups have been distinguished and named. The general term "tribe" is difficult to define because ethnic groups may be based on various and shifting constellations of shared language and culture, traditions of common ancestry, and more transient political factors. Relations among the many groups have stabilized since national political order was established in the mid-1960s, clan, tribal, and regional identities are important in all aspects of national life. Precise statistics do not exist, but it is unlikely that any group accounts for more than 10 percent of the total population. The largest group, the Kongo, may include as many as 2.5 million persons. Other socially and numerically important groups are the Luba, the Lunda, the Bashi, and the Mongo. Some groups, including the aboriginal Pygmies, occupy isolated ecological niches and number only a few thousand.

The Belgians introduced French, which now is spoken throughout the country by the educated. About 700 local languages and dialects also are spoken; the following four serve as official languages:

- Lingala developed along the Congo River in the 1880s in response to the need for a common commercial language. The Belgians made it the official language of the colonial armed forces. It was given a written form, and is now used extensively along the Zaire River from Kinshasa to Kinsangani and in the north and northwest.
- Swahili was introduced into the country by Arabs, especially the Zanzibari Swahilis, during the nineteenth century slaving operations. Swahili is spoken extensively in the eastern half of the country. In southern Shaba, a mutually intelligible variant called Kingwana is prevalent.
- Kikongo is used primarily between Kinshasa and the Atlantic Ocean as well as in parts of Congo and Angola. A modified form is used as a common tongue in southern Bandundu. Most languages of western Zaire belong to the Kinkongo group.
- Tshiluba is spoken primarily by the tribal groups of south-central Zaire.

About 80 percent of the Zairian population is Christian, predominantly Roman Catholic. Most of the non-Christians belong to either traditional religions or syncretic sects. Traditional religions embody such concepts as monotheism, animism, vitalism, spirit and ancestor worship, witchcraft, and sorcery, and vary widely among ethnic groups; none is formalized. The syncretic sects often merge Christianity with traditional beliefs and rituals. The most popular of these sects, Kimbanguism, was seen as a threat to the colonial regime and was banned by the Belgians. Kimbanguism, officially the "Church of Christ on Earth by the Prophet Simon Kimbangu," now has about 3 million members, primarily among the Bakongo of Bas Zaire and Kinshasa. In 1969, it was the first independent African church to be admitted to the World Council of Churches.

Before independence, education was largely in the hands of religious groups. The primary school system was well developed at independence; the secondary school system, however, was limited, and higher education was almost nonexistent. The principal objective of this system was to train low-level administrators and clerks. Since independence, efforts have been made to increase access to education, and secondary and higher education have been made available to many more Zairians. In 1980, about 80 percent of the males and 55 percent of the females, ages 6 to 11, were enrolled in a mixture of state- and church-operated primary schools. The 12 to 17 age group contains about twice as many male as female students. An estimated 30,000 students attend the national university and several technical and teacher-training institutes. The literacy rate among adult Zairians is 55 percent.

Sources: Haub and Yanagishita (1994); U.S. Department of State (1988).

Chapter References

Fox, Renee C. 1989. *The Sociology of Medicine: A Participant Observer's View*. Englewood Cliffs, NJ: Prentice-Hall.

Gallagher, Hugh. 1992. *NPR* "Morning Edition." (July 3).

Haub, Carl and Machiko Yanagishita. 1994. *1994 World Population Data Sheet*. Washington, DC: Population Reference Bureau.

Kelly, Michael. 1993. "David Gergen, Master of the Game." *The New York Times Magazine* (October 31):62-70+.

Rothman, David J. and Sheila M. Rothman. 1990. "How AIDS Came to Romania." *The New York Review of Books* (November 8):5-7.

______. 1993. "The New Romania." *The New York Review of Books* (September 23):56-57.

Stolberg, Sheryl. 1994. "Doctors' Dilemma." *Los Angeles Times* (April 5):E1+.

U.S. Department of State. 1988. "Zaire" *Background Notes* (#7793). Washington, DC: U.S Government Printing Office.

Wolff, Craig. 1992. "Alone, Undercover and Black: Hazards of Mistaken Identity." *The New York Times* (November 22):Y1+.

Answers

Concept Application

1. Stigmas
2. Role strain
3. Backstage; Stigmas
4. Sick role; Rights
5. Impression management

Multiple Choice

1.b 2.a 3.c 4.c 5.d 6.b 7.b 8.d 9.d 10.d 11.a 12.c 13.b 14.d 15.c 16.b 17.c 18.d 19.d 20.a 21.b 22.d 23.c 24.b 25.c 26.b 27.b 28.b 29.b 30.c 31.d

True/False

1.F 2.T 3.T 4.F 5.T 6.T 7.F 8.T 9.F 10.F 11.F 12.F 13.F

Chapter 7
Social Organization

Study Questions

1. What is the dominant theme of most disaster investigations?

2. What is an organization? How do sociologists approach the study of organizations?

3. Define multinational corporation. Give an example.

4. In what ways are multinational corporations engines of progress? In what ways are they engines of destruction?

5. What do sociologists mean when they say that organizations have two faces or two sides?

6. Why was Max Weber especially concerned with value-rational actions?

7. Define rationalization. Give an example. What are the positive and negative outcomes of rationalization?

8. Under what conditions would a prospective technology be considered optimum?

9. Use the concepts of value-rational action and externality costs to assess chemical companies in India.

10. What is a bureaucracy? How is studying a bureaucracy as an ideal type useful?

11. Distinguish between formal and informal dimensions of organizations. What are the positive and negative consequences associated with informal norms?

12. Explain how informal norms are related to the Bhopal crisis.

13. What is trained incapacity? Give an example, from Shoshana Zuboff's *In the Age of the Smart Machine*, of a work environment that promotes trained incapacity. Contrast that work environment with one that promotes empowering behavior.

14. What role did trained incapacity play in the Bhopal crisis? Is trained incapacity a problem unique to so-called underdeveloped countries like India?

15. How do organizations use statistical measures of performance? What are some of the problems that can accompany such measures? What role did statistical measures of performance play in the Bhopal crisis?

16. What is expert power? How can expert power be problematic?

17. Define oligarchy. Why does oligarchy seem to be an inevitable feature of large organizations?

18. How did Karl Marx define alienation? What is the source of alienation in the workplace? How do workers experience alienation?

19. How did alienation figure into the Bhopal crisis?

20. According to economist Robert Reich, which kinds of criteria should people use to judge the benefits of multinational corporations?

21. How does Weber's concept of value-rational action apply to the production of nuclear weapons during the Cold War?

22. Use the case of Denmark to support the argument that organizations can be agents of responsible action.

Concept Application

Below are five scenarios and the sources from which they were drawn. Decide which concept or concepts covered in Chapter 7 are represented best by each scenario, and explain why. The following concepts are considered:

Alienation
Bureaucracy
Disenchantment (of the world)
Efficiency seekers
Externality costs
Formal organization

Informal organization
Market seekers
Multinational corporation
Oligarchy
Organization
Professionalization
Rationalization
Resource seekers
Trained incapacity

Scenario 1

"Is IBM Japan an American or a Japanese company? Its work force of 20,000 is Japanese, but its equity holders are American. Even so, over the past decade IBM Japan has provided, on average, three times more tax revenue to the Japanese government than has Fujitsu. What is its nationality? Or what about Honda's operation in Ohio? Or Texas Instruments' memory-chip activities in Japan? Are they 'American' products? If so, what about the cellular phones sold in Tokyo that contain components made in the United States by American workers who are employed by the U.S. division of a Japanese company? Sony has facilities in Dotham, Alabama, from which it sends audiotapes and videotapes to Europe. What is the nationality of these products or of the operation that makes them?" (Ohmae 1990, p. 10).

Scenario 2

"Hospitals with hundreds, even thousands of inpatients maintain schedules aimed at ensuring that every patient receives essential care, and the staff must fit the needs and daily activities of dying patients into the hospital's schedule. They tend to require all patients, whether terminal or not, to give up virtually all personal control over the little things that make up their day-to-day lives. The kinds of personal items that can make a big difference, such as your own pillow from home, are often not allowed. Visits by children may be curtailed, and having a pet stay with a dying person is prohibited. Activities such as walking, eating, bathing, and any physical exercise will proceed according to an established routine" (Anderson 1991, p. 144).

Scenario 3

"A number of employees (5%) respond to perceived injustices by not performing their required tasks. One incident involved a male stockroom worker at a retail store who claimed he was paid less than others in similar positions. After an unsuccessful attempt to discuss the matter with his supervisor, he decided to deal with the conflict in his own way: 'I didn't really want to quit so I goofed off a lot. I didn't do anything unless I was specifically asked to. When working at night I would listen to music for hours and do nothing. . . .If I was goofing off and saw the manager, I would act as if I was really doing something' " (Tucker 1993, p. 37).

Scenario 4

"Scientifically the atomic bomb was an advance into unknown territory, but militarily it was simply a more cost-effective way of attaining a goal that was already a central part of strategy: a means of producing the results achieved at Hamburg and Dresden cheaply and reliably every time the weapon was used [for example, a quarter million bombs were used to destroy the city of Dresden]. (Even at the time, the $2 billion cost of the Manhattan Project was dwarfed by the cost of trying to destroy cities the hard way, using conventional bombs)" (Dyer 1985, p. 96).

Scenario 5

"The same kind of computer technology that enables employers to keep track of workers' backgrounds also makes it possible for them to quantify and monitor work performance. Anyone who works on a video display terminal, electronic telephone console, or other computer-based equipment, including laser scanner cash registers, is subject to constant monitoring.

Although the stated aim of monitoring workers is to improve productivity and service, the effect can be to turn checkstands into pressure cookers.

'Computers are wonderful for many things,' says Beverly Crownover, president of Local 1532 of United Food and Commercial Workers in Santa Rosa, California. 'But when they're used to monitor how many items a cashier scans per minute, it's like a whip. There's incredible pressure on workers' " (*UFCW Action* 1993, p. 135).

Applied Research

Select one of the world's top 25 largest industrial corporations listed in your textbook (see Table 7.3). Find out how the one you selected reaches around the globe through joint ventures, supply deals for parts and components, assembly operations, and marketing and distribution offices.

Practice Test

Multiple-Choice Questions

1. The dominant theme of most disaster investigations is that
   a. the technology failed in some way.
   b. workers at every level ignore, do not receive, fail to act on, fail to enforce, or fail to pass along information that could have prevented the disaster.
   c. one individual made a critical mistake.
   d. a unique set of factors caused the catastrophe.

2. From a sociological point of view, an organization is
   a. a body of persons legally recognized as having a purpose for meeting.
   b. a coordinating mechanism created by people to achieve stated objectives.
   c. a business enterprise.
   d. an official group of people who organize to make money.

3. From a sociological perspective, organizations
   a. cannot be studied apart from the people who create them.
   b. have a life that depends on the people who belong to them.

c. continue to exist even as their members die, quit, or return.
d. are coordinating mechanisms without clear objectives.

4. Max Weber maintained that the sociologist's main task is
   a. to explain solidarity.
   b. to analyze the causes and consequences of social action.
   c. to explain alienation.
   d. to focus on the question "How is social order possible?"

5. Weber was particularly concerned about the unforeseen consequences of ________ action.
   a. traditional
   b. affective
   c. value-rational
   d. instrumental

6. Weber defined rationalization as a process whereby thought and action rooted in emotion are replaced by thought and action rooted in
   a. logical assessment of cause and effect.
   b. mysterious forces.
   c. tradition.
   d. instrumental action.

7. The Bonda of India believe that spirits live in plants and trees. Therefore Bonda priests specify which trees can or cannot be cut down. According to Weber's typology of social action, this behavior toward trees represents an example of ________ action.
   a. affective
   b. value-rational
   c. instrumental
   d. planned

8. ________ action is the efficient application of means to some valued end.
   a. Traditional
   b. Affective
   c. Value-rational
   d. Instrumental

9. People living in a rationally organized environment typically know little about their surroundings and come to rely on ________ when something goes wrong.
   a. religion
   b. specialists
   c. their instincts
   d. chance

10. Max Weber maintains that one major side effect of rationalization is
   a. alienation.
   b. superstition.
   c. value-rational action.
   d. disenchantment of the world.

11. Costs above the price of producing a product, such as the cost of restoring land that has become contaminated are
   a. externality costs.
   b. social costs.
   c. environmental costs.
   d. secret costs.

12. The Cheverolet Corsica comes in two models; one accelerates more slowly but is more fuel-efficient than the other. When given a choice, most consumers prefer
   a. the faster, less fuel-efficient model.
   b. the slower, more fuel-efficient model.

13. Theoretically, in a bureaucracy
   a. authority belongs to the person.
   b. positions are filled on the basis of connections.
   c. authority resides in the personalities of people holding important positions.
   d. personnel treat clients as cases and without emotion.

14. The ________ dimensions of organizations consist of worker-generated norms that do not correspond to official policies, rules, and regulations.
   a. formal
   b. informal
   c. bureaucratic
   d. ideal

15. Worker-generated norms that govern output or physical effort are part of the
    a. informal dimension of organizations.
    b. formal dimension of organizations.
    c. rules and policies that define the goals of the organization.
    d. coordinating organizations.

16. When Marie (the personnel director) interviews Tran, a Vietnamese by birth, she finds that he does not say much about himself. Tran's reluctance to speak about himself can be traced to
    a. a lack of self-esteem.
    b. a poor grasp of the English language.
    c. a lack of education.
    d. Asian norms about how to present oneself during an employment interview.

17. At the Bhopal plant, workers monitored chemical leaks according to
    a. readings on gauges that monitored chemical pressure inaccurately.
    b. whether their eyes watered or burned.
    c. smell.
    d. formal regulations.

18. Trained incapacity is
    a. an inability to recognize the informal rules governing behavior.
    b. an ability to respond to unusual circumstances.
    c. an inability to recognize when official rules and procedures are outmoded.
    d. an ability to anticipate "what-if" scenarios.

19. A worker says, "Sometimes, I am amazed when I realize that we stare at the screen even when it has gone down." This comment suggests that in that organization, computers are used as
    a. an automating tool.
    b. an informating tool.
    c. a coordinating mechanism.
    d. a technological resource.

20. Blau and Schoenherr maintain that experts
    a. are trained by the organization for which they work.
    b. receive their training in colleges and universities.

c. are subjected to direct supervision.
d. are not permitted to be self-directed.

21. Which of the following is not one of the problems that can arise from using statistical measures of performance?
    a. The chosen measure may not be a valid indicator of what it is intended to measure.
    b. The measure may encourage workers to concentrate on achieving 'good scores' and to ignore problems generated by the desire to achieve a good score.
    c. Workers tend to pay attention only to those areas which are being measured and to overlook those for which no measure exists.
    d. Keeping records of such measures generates unnecessary paperwork.

22. When decision-making power is concentrated in the hands of a few persons who hold the top positions in an organizational heirarchy, the result is a state of
    a. oligarchy.
    b. alienation.
    c. disenchantment.
    d. trained capacity.

23. Although an emergency alarm sounded at the Bhopal plant, most people who lived around the plant did not know what it meant because
    a. they were illiterate peasants.
    b. the alarms sounded at least 20 times a week.
    c. they had never heard the alarm sound before.
    d. the alarms could not be heard above the fire sirens.

24. __________ is a state in which human life is dominated by the forces of human invention.
    a. Trained incapacity
    b. Alienation
    c. Oligarchy
    d. Professionalism

25. Community officials who recruit corporations to set up operations in their community often fail to consider
    a. the nationality of the corporation they are recruiting.
    b. the number of jobs the corporation is bringing to the commuinity.

c. the quality of jobs that the corporation is bringing to the community.
d. that American corporations should be given preference over a foreign corporation.

26. The Hanford Nuclear Reservation produced
   a. electrical power for a million people.
   b. plutonium to be used in atomic bombs.
   c. special containers to store nuclear materials.
   d. medicines that countered the effects of exposure to high doses of radiation.

27. The primary reason why managers of several industrial plants in Kalundborg, Denmark, have implemented programs to decrease pollution and waste is
   a. a desire to increase profit.
   b. concern about the environment.
   c. fear that the ozone layer will continue to deteriorate.
   d. a local government mandate.

28. The "value" governing the Hanford Nuclear Reservation's goal of producing plutonium for nuclear weapons was
   a. to find the safest means of producing plutonium.
   b. to keep bomb production operating 24 hours a day.
   c. to win the arms race against the Soviet Union in order to deter communism.
   d. to manufacture 60,000 nuclear warheads at a cost of $750 billion.

True/False Questions

T F 1. Industrial accidents have occurred in the United States, in which the chemicals released exceeded those released in Bhopal in amount and toxicity.

T F 2. Blau and Schoenherr maintain that experts hired by an organization have little control over the application of the information, service, or invention they provide to the organizaition.

T F 3. Hazardous chemicals banned in the United States cannot be manufactured by American chemical corporations with production facilities in foreign countries.

T F 4. Multinational corporations have headquarters disproportionately in the United States, Japan, and western Europe.

T F 5. Multinational corporations plan, produce, and sell on a national scale.

T F 6. Rationalization discredits the idea that plants and trees have spirits.

T F 7. To informate means to use the computer as a source of surveillance.

T F 8. Trained incapacity is a problem unique to underdeveloped countries such as India because a lack of "what-if" thinking is rooted in the culture of that country.

T F 9. Residents of Institute, West Virginia, responded to a chemical leak at the local Union Carbide plant in the same disorganized way as Bhopal residents.

Continuing Education

Leisure

For a change in music, listen to

The Sounds of India (Ravi Shankar). Columbia.
Lower Caste Religious Music. Lyricord.
Miracle Percussion of Kutyattam: The Oldest Drama in the World. JVC (Ethnic Sound Series).

The next time you rent a movie, consider

Silkwood, directed by Mike Nichols (131 minutes). The story is based on the true story of Karen Silkwood, a peace activist and an employee of a nuclear plant, who died in a car accident on her way to testify about the dangerous and questionable practices at the plant. As you view the film, use the sociological concepts presented in this chapter to think about the factors contributing to the existence of such dangerous working conditions. Also, consider how workers responded to these conditions and how their work-related stresses affected their personal relationships at and outside the workplace.

Travel

If you're in Atlanta, visit The World of Coca-Cola Pavilion (Martin Luther King, Jr. Drive at Central Avenue; 404-676-5151). The pavilion contains Coca-Cola artifacts and memorabilia that explores the more than a 100-year history of this well-known global corporation.

If you're in New York City, visit the AT&T Infoquest (550 Madison Avenue at E. 56th Street, 212-605-5555). Approximately 40 interactive exhibits permit people to "experience" various communication and information technologies and to gain insights about the services that this global corporation markets to the world.

General and Readable Information

India Today. Living Media India Ltd., 404 S. Park Avenue, New York 10016. The format is similar to that of *Time*, *Newsweek*, and *U.S. News & World Report* except that this is a bimonthly publication. The magazine provides a broad overview of current events in India as well as international events that are of interest to Indian readers.

Country Profile
India

Official Name: Republic of India
Population (1994 est.): 911,600,000
Land Area: 1,147,950 square miles
Population Density: 794 people per square mile
Birth Rate per 1,000: 29
Death Rate per 1,000: 10
Rate of Natural Increase: 1.9
Average Number of Children Born to a Woman During Her Lifetime: 3.6
Infant Mortality Rate: 79 deaths per 1,000 live births
Life Expectancy at Birth: 57 years
Gross National Product per Capita: $310
Capital: New Delhi

Although India occupies only 2.4 percent of the world's land area, it supports nearly 15 percent of the world's population. Only China has a larger population. A large percentage of India's population is in its teens; 40 percent of Indians are younger than age 15. About 80 percent of the people live in more than 550,000 villages, and the remainder in more than 200 towns and cities.

Northern India has been invaded from the Iranian plateau, Central Asia, Arabia, and Afghanistan at various times in its ancient and pre-modern history. The blood and culture of these invaders have mixed freely with that of the indigenous people, contributing to an unparalleled degree of racial and cultural synthesis. Religion, caste, and language are major determinants of social and political organization in India today. Sixteen officially recognized languages are spoken in India; Hindi is spoken most widely.

Although 83 percent of the people are Hindu, India also is the home of more than 80 million Muslims, giving it one of the world's largest Muslim populations. Adherents to other religions include Christians, Jews, Sikhs, Jains, Buddhists, and Parsis.

The caste system, comprising the traditional social divisions of Indian society, has been historically based on occupation-related categories ranked in a theoretically defined hierarchy. Traditionally, four castes were identified, plus a category of outcastes or untouchables. In reality, however, there are thousands of subcastes and it is with these subcastes that the majority of Hindus identify. Despite economic modernization and laws countering discrimination at the lower end of the class structure, the caste system remains an important factor in Indian society.

Sources: Haub and Yanagishita (1994); U.S. Department of State (1989).

Chapter References

Anderson, Patricia. 1991. *Affairs in Order: A Complete Resource Guide to Death and Dying*. New York: Macmillan.

Dyer, Gwynne. 1985. *War*. New York: Crown.

Haub, Carl and Machiko Yanagishita. 1994. *1994 World Population Data Sheet*. Washington, DC: Population Reference Bureau.

Ohmae, Kenichi. 1990. *The Borderless World: Power and Strategy in the Interlinked Economy*. New York: Harper Business.

Tucker, James. 1993. "Everyday Forms of Employee Resistance." *Sociological Forum* 8(1):25-45.

UFCW Action. 1993. "The Boss Is Watching." *Utne Reader* (May/June):134-35.

U.S. Department of State. 1989. "India." *Background Notes* (#7847). Washington, DC: U.S. Government Printing Office.

Answers

Concept Application

1. Multinational corporation
2. Bureaucracy, Formal organization
3. Informal dimension of organizations
4. Rationalization
5. Automate

Multiple-Choice

1.b 2.b 3.a 4.b 5.c 6.a 7.a 8.c 9.b 10.d 11.a 12.a
13.d 14.b 15.a 16.d 17.b 18.c 19.a 20.b 21.d 22.a 23.b
24.b 25.c 26.b 27.a 28.c

True/False

1.T 2.T 3.F 4.T 5.F 6.T 7.F 8.F 9.T

Chapter 8
Deviance, Conformity, and Social Control

Study Questions

1. What is deviance? How is it related to conformity and social control?

2. Is it possible to generate a list of deviant behavior? Why or why not?

3. What is the unique contribution of sociology to the study of deviance? What two fundamental questions guide the sociological perspective on deviance?

4. Why is China the country of emphasis in a chapter on deviance?

5. Describe how ideological commitment is connected to issues of deviance, conformity, and mechanisms of social control in China.

6. Briefly describe the Cultural Revolution. Why was making money a punishable crime during the Cultural Revolution but not after Mao's death?

7. Distinguish between folkways and mores. Give an example of Chinese and American mores.

8. What important cultural lessons are incorporated into the daily activities of Chinese and American preschoolers?

9. What are the major mechanisms of social control? Why do all societies have such mechanisms in place?

10. Briefly explain the following statement: "In China social control is everywhere and involves everyone."

11. According to Durkheim, why is crime a "normal" and necessary phenomenon?

12. Labeling theorists believe that rules are socially constructed and that members of social groups do not enforce them in uniform or consistent ways. Explain.

13. What are the implications of the categories "secret deviant" and "falsely accused" for the study of deviance?

14. Under which circumstances are people most likely to be falsely accused of a crime?

15. What are witch hunts? Why do they occur? Give an example.

16. What is white-collar crime? Why are white-collar criminals less likely to be caught than "common" criminals?

17. Compare and contrast Chinese and American conceptions of guilt or innocence and of the role of lawyers.

18. Who are claims makers? What factors determine a claim maker's success?

19. Describe the constructionist approach to analyzing claims makers and claimsmaking activities.

20. Explain Stephen J. Gould's position on the distinction between a legal substance such as tobacco and an illegal substance such as cocaine.

21. What is structural strain? What are the sources of structural strain in the United States?

22. What are the responses to structural strain?

23. Identify one source of structural strain in China. Use Merton's typology of response to consider how people in China respond to this strain.

24. Summarize the major assumptions underlying the theory of differential association.

25. How does the assumption underlying the theory of differential association relate to mechanisms of social control in China?

26. What larger historical and geographical factors explain the differences between the Chinese and the American systems of social control?

27. Why is deviance a complex concept?

28. In Milgram's classic experiment *Obedience to Authority*, why did a significant number of volunteers come to accept an authority's definition of deviance and administer shocks although they caused obvious harm to confederates?

Concept Application

Below are five scenarios and the sources from which they are drawn. Decide which concept or concepts covered in Chapter 8 are represented best by each scenario, and explain why. The following concepts are considered:

Claims makers
Corporate crime
Deviant subcultures
Differential association
Falsely accused
Formal sanctions
Informal sanctions
Mechanisms of social control
Sanctions
Secret deviants
White-collar crimes
Witch-hunt

Scenario 1

"The Tobacco Institute was founded in 1958, even before the first Surgeon General's report on the health risks of smoking, to represent the interests of tobacco companies to lawmakers. Once financed by a dozen companies, it now works for only five--Philip Morris, R.J. Reynolds, Lorillard, Liggett and American Brands--but its twofold mission remains the same: to persuade Federal, state and local authorities to lay off and to sell the virtues of the industry to the American public. A staff of lobbyists handles the first task and Ms. Dawson, at 32, the second.

The job description is fairly typical for a trade organization--to develop and articulate the industry position on any given issue, then make sure the message reaches the public. But this is no typical industry" (Janofsky 1994, p. 8F).

Scenario 2

"Boesky told the government about his insider trading activities, not only with me, but with at least one other well-known investment banker. Beyond that, he detailed various schemes, concocted with those in the highest circles of power, to circumvent SEC regulations and tax laws. Said Carroll, 'He has played fast and loose with the rules that govern our markets, with the effect of manipulating the outcome of financial transactions measured in the hundreds of millions of dollars' " (Levine and Hoffer 1991, p. 346).

Scenario 3

"The small-time criminals are everywhere. Maybe they're sneaking into more than one theatre in the local cineplex. Or grabbing a handful of yogurt peanuts from the grocery store bin and eating all the evidence before getting to the check-out stand. Or making personal long-distance calls from work" (Tomashoff 1993, p. E1).

Scenario 4

"Death sentences for people who later prove to be innocent are less unusual than is commonly supposed. Just in the last five months, four once condemned prisoners have been released after spending years on death row. Two of them, in Alabama and Texas, turned out to have been convicted on fabricated evidence and perjured testimony; the third, in Texas, was convicted because of evidence that was withheld; the fourth, in Maryland, was exonerated by DNA analysis, a technology that was unavailable at that time of his trial" (*The New Yorker* 1993, p. 4).

Scenario 5

Can a court force an unwilling person to give up part of his or her body (e.g., bone marrow, a kidney) to a relative who needs that body part to survive? That was the question recently brought before the court of common pleas in Allegheny County, PA. The common law has consistently held that one human being is under no legal obligation to give aid or take action to save another human being or to rescue one. The court said that such a rule, although revolting in a moral sense, is founded upon the very essence of a free society, and while other societies may view things differently, our society has as its first principle respect for the individual--and society and government exist to protect that individual from being invaded and hurt by another (Chayet 1983).

Applied Research

Almost every American viewed the amateur-produced videotape showing a group of white police officers beating Rodney King, a 25 year-old unarmed black man. Among other things, King suffered 11 skull fractures, some brain damage, a crushed cheekbone, and a number of internal injuries. *The New York Review of Books* published a two-part series of the report of the Independent Commission on the Los Angeles Police Department. The author, John Gregory Dunne, also reviewed a biography of Los Angeles Chief of Police Daryl Gates, published in 1982 by the *Los Angeles Times*, "Daryl Gates: A Portrait of Frustration."

This Independent Commission was charged with the task of investigating and identifying factors which contributed to this gross misconduct. Dunne's two reviews appeared in the October 10 and October 24, 1991 issues.

On the basis of the information presented in the reviews, state what you believe were the three foremost factors or conditions leading to the police misconduct that was displayed in the Rodney King incident.

Practice Test

Multiple-Choice Questions

1. The only characteristic common to all forms of deviance is the fact that
   a. they invoke formal sanctions.
   b. everyone in the society is offended by the behavior.
   c. the behaviors are considered deviant across time and place.
   d. some social audience regards them and treats them as deviant.

2. In answering the question, "How is it that almost any behavior or appearance can qualify as deviant under certain circumstances?" sociologists focus on
   a. the broader historical and social context.
   b. the deviant personality.
   c. social class.
   d. formal sanctions.

3. In China "profit-making" activities were considered criminal during the
   a. 1990s.
   b. 1980s and 1990s.
   c. 1960s and 1970s.
   d. 1940s.

4. Chinese students who participated at Tiananmen Square in support of democracy were accused of "bourgeois liberalization," or
   a. profit-making activities.
   b. wanton expression of individual freedom (individualism).
   c. selflessness.
   d. conspiracy.

5. Which of the following is a false statement about the Cultural Revolution?
   a. The Cultural Revolution was inspired by the Communist Party Chairman Mao Zedong.
   b. The Cultural Revolution was a campaign against revisionism and against entrenched authority.
   c. During the Cultural Revolution the Chinese sought to preserve and protect many artifacts from China's long history.

d. The Cultural Revolution was an attempt by Mao Zedong to eliminate anyone in the Communist Party and in the masses who opposed him.

6. The ________ was a plan of Mao's to mobilize the masses and transform China from a country of poverty to a land of agricultural abundance in five short years.
   a. Five Year Plan
   b. Great Leap Forward
   c. Cultural Revolution
   d. Special Economic Zones Project

7. Which one of the following does not apply to the concept of folkways?
   a. pivotal
   b. customary
   c. routine
   d. details of life

8. Mores in China are directed toward preserving
   a. individual rights.
   b. privacy.
   c. collective rights.
   d. status.

9. Chinese preschool teachers tend to discipline their four-year-olds by
   a. stopping them from misbehaving before they know they are about to misbehave.
   b. establishing the rules and enforcing them consistently whenever they are broken.
   c. rewarding them with small toys whenever they are good.
   d. punishing them harshly whenever they break a rule.

10. Informal sanctions are
   a. backed by the force of law.
   b. spontaneous and unofficial expressions of approval or disapproval.
   c. group-generated expressions of disapproval or approval.
   d. systematic laws, rules, and regulations.

11. In China "social control is everywhere and involves everyone." This means that
   a. secret police are everywhere.
   b. the Chinese people are encouraged to report wrongdoings.

c. everywhere there are plainclothes police watching everyone.
d. almost every Chinese person has spent some time in prison.

For questions 12-19 use one from the following set of responses to match each statement with the appropriate theory of deviance.

a. Functionalist perspective (as represented by Emile Durkheim)
b. Labeling theory
c. Differential association
d. Constructionist approach
e. Structural strain

12. A group that goes too long without noticing crime or doing something about it would lose its identity as a group.

13. Labels, examples, and orientation are important because they tend to evoke a particular cause and a particular solution for a social problem.

14. Deviance is likely to be high when the legitimate opportunities for meeting the culturally valued goals are closed to a significant portion of people.

15. Criminal behavior is learned.

16. Members of social groups do not enforce rules uniformly or consistently.

17. Crime is present in all societies.

18. Whether an act is deviant depends on whether people notice it and, if they do so, whether they subsequently apply sanctions.

19. Violating a rule does not make a person deviant.

20. ________ are people who have broken the rules and are caught, punished, and labeled as outsiders.
    a. Secret deviants
    b. The falsely accused
    c. Conformists
    d. Pure deviants

21. The internment of more than 110,000 people of Japanese descent in the United States during World War II is an example of
    a. a crime.
    b. an informal mechanism of social control.
    c. a structural strain.
    d. a witch-hunt.

22. In studying the Tiananmen Square incident, constructionists would focus on which one of the following questions?
    a. Who controlled the information that government troops received about the demonstration?
    b. Why is making money acceptable in China today?
    c. What sources of structural strain underlie the demonstration?
    d. What measures did the government take to punish student demonstrators?

23. Former Surgeon General Koop's hypothesis that cigarette smokers deprived of their drug would behave like cocaine and heroine addicts is supported by behavior of
    a. smokers in the United States when someone asks them not to smoke.
    b. millions of men and women in prison who have no access to cigarettes.
    c. medical doctors who smoke even though they understand the risks.
    d. 13 million smokers in Italy who were prevented from purchasing cigarettes because of a strike.

24. According to Robert K. Merton, considerable structural strain exists in the United States because
    a. opportunities are open to all.
    b. people must go to college in order to become successful.
    c. American culture places a high value on social advancement for all its members regardless of the circumstances into which they are born.
    d. the legitimate means to achieve the culturally valued goals are clearly defined.

For questions 25-27 use one from the following set of responses to decide which applies to each statement.

a. Conformity
b. Innovation
c. Ritualism

d. Retreatism
e. Rebellion

25. Success is equated with winning the game rather than with playing by the rules of the game.

26. One plays the game according to the rules even if one is defeated.

27. The person "resigns" from society.

28. According to Merton's typology of responses associated with structural strain, couples in China would be classified as retreatists if they
    a. decided to abort a baby because it was a girl.
    b. disagreed with the one-child policy.
    c. claimed ethnic minority status in order to have more than one child.
    d. hid the birth of baby girls from party officials.

29. The idea of ________ is the basis of the Chinese philosophy in which rehabilitation is regarded as the purpose of punishment.
    a. structural strain
    b. differential association
    c. labeling
    d. claims making

30. The habitable land area of China is approximately
    a. the size of the United States.
    b. the size of Europe, North America, and South America combined.
    c. half the size of the United States.
    d. the size of the African continent.

31. Which of the following is not one of the strong and persistent traits that has existed over China's 3,700-year-long history?
    a. A system of family responsibility
    b. The Confucian tradition
    c. Smooth transition of power from one regime to the next
    d. An imperial tradition

32. In research, a person working in cooperation with an experimenter is known as
    a. the control agent.
    b. a confederate.
    c. the experimented-on.
    d. a double agent.

33. In Stanley Milgram's classic experiment, *Obedience to Authority*, Miligram found that obedience was founded on
    a. the firm command of a person with a status that gave minimal authority over a subject recruited to participate in the study.
    b. the subject's fear of being punished physically if he or she disobeyed.
    c. the subject's dislike of the learner's physical characteristics.
    d. the subject's firm belief that learning is enhanced when feedback regarding failure is immediate.

True/False Questions

T F 1. Almost any behavior or appearance can qualify as deviant under the right circumstances.

T F 2. China has a trade surplus with the United States.

T F 3. China is the host country of the 2000 Olympic Games.

T F 4. Ideally, conformity is voluntary.

T F 5. According to Emile Durkheim, deviance will be present even in a community of saints.

T F 6. Deviance is a consequence of a particular behavior or appearance.

T F 7. Stephen J. Gould argues that alcohol and tobacco should be classified as illegal drugs.

T F 8. In the United States, prisons function to rehabilitate inmates.

T F 9. The Chinese believe in early intervention, before deviant behavior has become extreme or has caused too much damage.

Continuing Education

Leisure

For a change in music, listen to

Like Waves Against the Sands: (Jing Ying Ensemble) Saydisc.
Mongolian Folk Music. Hungar.
Sounds and Music of China. Monitor.

The next time you rent a movie, consider

Victim, a British film directed by Basil Dearden (1962, 100 minutes). This is one of the first films to consider the social and personal experiences of homosexuality in a serious way. The film is set in a time before homosexuals organized and encouraged gay men and women to "come out of the closet." As you view the film, consider how homosexuals' personal and social experiences created a need to organize and to demand that gay rights be recognized.

Travel

If you're in Kansas City, Missouri, visit the Nelson-Atkins Museum (4525 Oak Street; 816-561-4000). This museum houses an outstanding collection of Chinese art from all historical periods.

If you're in Seattle, visit the Wing Luke Asian Museum (407 Seventh Avenue S; 206-623-5124). This museum focuses on Chinese migration to the northwest since 1860. It also contains art and folklore exhibits representative of the Far East in general.

If you're in New York City, visit the Police Academy Museum (235 E. 20th Street; 212-477-9753). This museum is part of the New York City Police Academy, and exhibits 150 years of police artifacts and memorabilia.

General and Readable Information

China Now. Society of Chinese Understanding, 152 Camden High Street, London NW1. This illustrated quarterly magazine covers a broad range of topics about Chinese people; narratives, biographies, interviews, translations, fiction, society, and culture.

China Report. Washington Center for China Studies, P.O. Box 32069, Washington, DC 20007. These series of research papers focus on China's economy, politics, society, culture, and foreign relations.

Country Profile
China

Official Name: People's Republic of China
Population (1994 est.):1,192,000,000
Land Area: 3,600,930 square miles
Population Density: 331 people per square mile
Births Rate per 1,000: 18
Deaths Rate per 1,000: 7
Rate of Natural Increase: 1.1%
Average Number of Children Born to a Woman During Her Lifetime: 2.0
Infant Mortality Rate: 31 deaths per 1,000 live births
Life Expectancy at Birth: 70 years
Gross National Product per Capita: $380
Capital: Beijing

China's population in mid-1993 was about 1.4 billion, with an estimated growth rate of 1.2%. China is very concerned about its population growth and has attempted to implement a strict population control policy. The government's goal is one child per family, with exceptions in rural areas and for ethnic minorities. This policy is often ignored in the countryside and also by many urban dwellers. The government states that it opposes physical compulsion to submit to abortion or sterilization, but the instances of coercion have continued as local officials strive to meet population targets. The government's goal is to stabilize the population early in the 21st century, although some current projections estimate a population of 1.6 billion by 2025. Overall annual population growth is estimated to have dropped to about 1.2% since 1973.

The largest ethnic group is the Han Chinese, who constitute about 93% of the total population. The remaining 7% are Zhuang (16 million), Manchu (9 million), Hui (8 million), Miao (8 million), Uygur (7 million), Yi (7 million), Tibetan (5 million), Mongol (5 million), and Korean (1 million).

There are seven major Chinese dialects and many subdialects. The Beijing dialect, often called Mandarin (or Putonghua), is taught in all schools, and is the medium of government. Only about two-thirds of the Han ethnic group are native speakers of Mandarin; the rest, concentrated in southwest and southeast China, speak one of the six other major Chinese dialects. Non-Chinese languages spoken widely by ethnic minorities include Mongolian, Tibetan, Uygur and other Turkic languages in Xinjiang, and Korean in the northeast.

Religion continues to play a significant part in the life of many Chinese. Buddhism is most widely practiced, with an estimated 100 million adherents; traditional Daoism also is practiced. Official figures indicate there are 20 million Muslims, 3.6 million Catholics, and 5.6 million Protestants; unofficial estimates are much higher.

While the constitution affirms religious toleration and the Chinese Government has reopened many temples, mosques, and churches closed during the Cultural Revolution, the government places restrictions on religious practice outside officially recognized organizations. The government permits only two official Christian organizations, a Catholic church without ties to Rome and the "Three-Self Patriotic" Protestant church. Underground churches have sprung up in many parts of the country.

Sources: Haub and Yanagishita (1994); U.S. Department of State (1993).

Chapter References

Chayet, Neil. 1983. "Law and Morality." Pp. 418-19 in *Life Studies: A Thematic Reader*, edited by D. Cavitch. New York: St. Martin's.

Dunne, John Gregory. 1991. "Law and Disorder in Los Angeles." *The New York Review* (October 10):23-29.

______. 1991b. "Law and Disorder in Los Angeles." *The New York Review* (October 24):62-70.

Haub, Carl and MachikoYanagishita. 1994. *1944 World Population Data Sheet*. Washington, DC: Population Reference Bureau.

Janofsky, Michael. 1994. "Antismoking Forces at the Barricades? Bring 'em On!" *The New York Times* (April 24):8.

Levine, Dennis B. and William Hoffer. 1991. *Inside Out: An Insider's Account of Wall Street*. New York: Putnam.

Tomashoff, Craig. 1993. "America's Least Wanted Criminals." *Los Angeles Times* (May 10):E1+.

The New Yorker. 1993. "Wrongful Death." (August 16):4-6.

U.S. Department of State. 1993. "China." *Background Notes* (#7847). Washington, DC: Population Reference Bureau.

Answers

Concept Application

1. Claims maker
2. White-collar crime
3. Secret deviant
4. Falsely accused
5. Mores

Multiple-Choice

1.d 2.a 3.c 4.b 5.c 6.b 7.a 8.c 9.a 10.b 11.b 12.a 13.d 14.e 15.c 16.b 17.a 18.b 19.b 20.d 21.d 22.a 23.d 24.c 25.b 26.c 27.d 28.d 29.b 30.c 31.c 32.b 33.a

True/False

1.T 2.T 3.F 4.T 5.T 6.F 7.F 8.F 9.T

Chapter 9
Social Stratification

Study Questions

1. What is the connection between social stratification and life chances?

2. Why was South Africa chosen as the country to emphasize in reference to issues of social stratification?

3. What two major kinds of criteria do societies use to categorize people?

4. What does status value mean?

5. Explain the rationale used by Jane Elliot to divide her class according to eye color.

6. What is the major shortcoming of all classification schemes? How do people tend to deal with this shortcoming?

7. In a purely biological sense, what is race? What are some of the shortcomings associated with any racial classification scheme?

8. What characteristics distinguish a caste from a class system of stratification?

9. Explain the basic dynamics of apartheid.

10. Is the United States a class system? Why or why not?

11. Are caste and class systems distinct types of stratification systems? Explain.

12. How do functionalists (Davis and Moore) explain stratification? What are some of the shortcomings of their explanations?

13. Summarize how Marx approached social class in his writings. Identify three ideas that Marx gave us for approaching social class.

14. How does Max Weber use the concept of social class?

15. How is class ranking complicated by status groups and parties?

16. What general structural changes in the American economy have created an underclass?

17. What are the broad characteristics of social mobility in the United States?

Concept Application

Below are five scenarios and the sources from which they were drawn. Decide which concept or concepts covered in Chapter 9 are represented best by each scenario, and explain why. The following concepts are considered:

Achieved characteristics
Apartheid
Ascribed characteristics
Caste system
Class system
Downward mobility
Intergenerational mobility
Intragenerational mobility
Life chances

Race
Social stratification
Status group
Status value
Upward mobility
Urban underclass
Vertical mobility

Scenario 1

"Do blondes have more fun? Social scientists have yet to nail down the answer. But economists now have good reason to believe that blondes make more money--or at least the trim, attractive ones do. New studies show that men and women (with any hair color) who are rated below average in attractiveness by survey interviewers typically earn 10 to 20 percent less than those rated above average.

One is tempted to write off the results as proof that idle econometricians are the Devil's helpers. But the findings from Daniel Hamermesh of the University of Texas and Jeff Biddle of Michigan State are complemented by other research showing that obese women are also at a considerable earnings disadvantage. And they could figure prominently in the very serious business of deciding who is protected by the three-year-old Americans with Disabilities Act" (Passell 1994, p. C2).

Scenario 2

"It was all part of a vast national phenomenon. The number of families moving into the middle class--that is, families with more than five thousand dollars in annual earnings after taxes--was increasing at the rate of 1.1 million a year, *Fortune* noted. By the end of 1956 there were 16.6 million such families in the country, and by 1959, in the rather cautious projections of *Fortune*'s editors, there would be 20 million such families--virtually half the families in America. *Fortune* hailed 'an economy of abundance' never seen before in any country in the world. It reflected a world of 'optimistic philoprogenitive (the word means that Americans were having a lot of children) high-spending, debt-happy, bargain-conscious, upgrading, American consumers' " (Halberstam 1993, p.587).

Scenario 3

"The Brinks Hotel was another American symbol in Saigon. It was a bachelor officers' headquarters, an American world that Vietnamese need not enter unless of course it was to clean the rooms or to cook, or to provide some other form of service. It stood high over Saigon and its poverty and its hovels, a world of Americans eating American food, watching American movies, and just to make sure that there was a sense of home, on the roof terrace there was always a great charcoal grill on which to barbecue thick American steaks flown in especially to that end" (Halberstam 1987, Pp. 618-19).

Scenario 4

"Whenever more than a few people come together to work on a task, a hierarchy is established that leads to accepted norms concerning who is considered the leader and/or who can give and who should accept directions....The criteria for placement in the hierarchy vary from culture to culture and from task to task within any one culture. The criteria might include age, birthright, election by peers, expertise in a topic area, family name, formal education, sex, or even physical attractiveness. No matter what exact terms are used or what the exact criteria are, some people within a culture are said to be at the top of a hierarchy and consequently possess more status. These people have certain rights, such as deference from others, the right to speak first at a meeting, the expectation that others will accept directions, and the expectation that their opinions will affect decisions" (Brislin, Cushner, Cherrie, and Yong, 1990, p. 293).

Scenario 5

"These children [of people who make enough money to live a privileged life] learn to live with *choices*: more clothes, a wider range of food, a greater number of games and toys, that other boys and girls may ever be able to imagine. They learn to grow fond of, or resolutely ignore, dolls and more dolls, large dollhouses and all sorts of utensils and furniture to go in them, enough Lego sets to build yet another house for the adults in the family. They learn to take for granted enormous playrooms, filled to the brim with trains, helicopters, boats, punching bags, Monopoly sets.... They learn to assume instruction--not only at school, but at home--for tennis, for swimming, for dancing, for horse riding. And they learn often enough to feel competent at those sports, in control of themselves while playing them, and, not least, able to move smoothly from one to the other" (Coles 1978, p. 26).

Applied Research

In *Images of Japanese Society*, Ross E. Mouer and Yoshio Sugimoto (1990) present a multidimensional framework for thinking about stratification. They identify four dimensions of stratification: economic, political, psychological, and information-based, and give examples of rewards associated with each dimension. Select one specific type of reward from the list below and find data that demonstrate patterns of inequality characteristic of American society.

Economic rewards:

- Occupation
- Salary
- Pension
- Benefits
- Environment (quality of surroundings)
- Employment security
- Job safety
- Quality of recreation facilities
- Leisure

Political rewards:

- Influence
- Authority
- Contacts
- Access to guns and tanks
- Control over army or police force
- Votes

Publicity
Information and intelligence

Psychological rewards:
Status
Prestige
Honor
Esteem
Fame
Publicity
Recognition
Friends
Conspicuous consumption

Information-based rewards:
Knowledge
Specific skills
Social awareness
Technical know-how
Access to information

Practice Test

Multiple-Choice Questions

1. Social stratification is _______ process in which people in societies are ranked on a scale of social worth.
   a. a random
   b. an arbitrary
   c. a systematic
   d. an automatic

2. Sociologists study how the categories in which people are placed effect their
   a. classification.
   b. self-esteem.

c. ascribed status.
d. life chances.

3. If a country is to establish a true democracy,
   a. its system of social stratification must be "democratic."
   b. its elections must be free and fair.
   c. its government must be multiracial.

4. Sociologists are particularly interested in situations in which _________ are assigned different status value.
   a. achieved characteristics
   b. eye colors
   c. roles
   d. ascribed characteristics

5. Jane Elliot's classic experiment, in which she divided the pupils in class according to eye color, shows how easy it is for people to
   a. assign social worth to achieved characteristics.
   b. explain behavior in terms of an ascribed characteristic.
   c. build a reward system around a very significant attribute.
   d. classify people when the categories are clear-cut.

6. The "pencil" test and the "pine slab" test show that
   a. racial categories are clear-cut.
   b. sharp dividing lines distinguish white skin from brown skin.
   c. a lack of clear racial categories does not keep people from trying to create them.
   d. physical traits such as hair texture and skin color can be used to distinguish different races.

7. According to the rule of hypodescent, descendants of nonwhite-white unions in the United States are
   a. classified as white.
   b. assigned to a special racial category.
   c. assigned to a mixed-race category.
   d. classified as members of the nonwhite group.

8. Caste systems and class systems are ideal types. This means that they represent
   a. real cases of social stratification.

b. stratification systems with desirable characteristics.
c. two standards against which real cases can be compared.
d. South Africa and the United States respectively.

9. The Lands Act reserved 85 percent of South African land for approximately _______ percent of the population.
a. 15
b. 30
c. 50
d. 60

10. Before the apartheid laws were dismantled, African people were declared
a. citizens of Africa.
b. citizens of one of 10 homelands.
c. noncitizens.
d. citizens of South Africa.

11. Whites in South Africa consist of two distinct types:
a. Afrikaners and English speakers.
b. Boers and Afrikaners.
c. South African whites and Dutch speakers.
d. Boers and Dutch.

12. Apartheid is a classic example of _______ system of stratification.
a. an intragenerational
b. a downward
c. a class
d. a caste

13. In a true class system there is
a. a systematic connection between ascribed characteristics and social class.
b. no systematic connection between ascribed characteristics and social class.
c. complete equality (a plumber has the same status as a judge).
d. no system of social stratification.

14. _______ mobility is a gain or loss of rank, as when an accountant becomes unemployed.
a. Vertical

b. Intergenerational
c. Horizontal
d. Social

15. Most systems of stratification, including the U.S. system, are
    a. class systems.
    b. caste systems.
    c. a combination of class and caste systems.
    d. intragenerational class systems.

16. In which one of the following occupational categories are women most likely to be overrepresented?
    a. Material-moving occupations
    b. Dental assisting
    c. Bookkeeping
    d. Fire prevention

17. As of 1990, 25 percent of professional baseball players were black or Hispanic. If the on-field positions were filled without regard to ethnicity, _______ of pitchers should be black or Hispanic.
    a. 25 percent
    b. approximately 3 percent (25 divided by 9 positions)
    c. 10 percent
    d. 50 percent (because pitchers make up half the team)

18. In which of the following occupational categories are people of Hispanic origin most likely to be overrepresented?
    a. Nurses' aides
    b. Farmworkers
    c. Waiters and waitresses
    d. Technical writers

19. According to the functionalist perspective, the unequal distribution of rewards is necessary in order to
    a. ensure that the most functionally important occupations are filled by the best-qualified people.
    b. make the least functionally important occupations attractive to the masses.

c. justify denying some people the opportunity to achieve functionally important occupations.
d. make the system as democratic as possible.

20. Critics of the functional perspective on social stratification argue that this approach falls short because
    a. one must assume that social stratification exists in all societies.
    b. workers who perform the same jobs tend to receive equal pay regardless of their race or sex.
    c. salary reflects an occupation's contribution to society.
    d. it is difficult to determine the functional importance of an occupation.

21. Marx believed that conflict between _______ distinct social classes propels society from one historical epoch to another.
    a. six
    b. four
    c. three
    d. two

22. Which of the following is not one of the Marxist ideas about social class?
    a. Conflict between two distinct classes propels society from one historical epoch to another.
    b. Classes can be distinguished from one another according to sources of income.
    c. Class ranking is complicated by status groups and parties.
    d. The success or failure of class conflict depends on many factors, including the support of other classes.

23. Weber's ideas about social class inspires sociologists to
    a. study the people who make up the middle class.
    b. compare the situation of the wealthiest with that of the very poor.
    c. study class conflict as an agent of change.
    d. think of social class as determined by wealth and income.

24. Which one of the following characteristics does not fit the profile of the urban underclass?
    a. Homogeneous
    b. At the very bottom of the economic hierarchy

c. Consisting of families and individuals
d. Inner-city

25. South Africa's attempt to truly dismantle apartheid will hinge on
    a. who becomes president in 1998.
    b. whether it can dismantle the legacy of this policy.
    c. whether the new government abolishes the laws of apartheid.
    d. whether blacks assimilate into the white population.

26. Although it is difficult to make a precise statement about the characteristics of mobility in the United States, we can say that
    a. there is a considerable amount of rags-to-riches mobility.
    b. a child whose father is in the bottom 5 percent of earners has no chance of mobility.
    c. going from rags-to-riches or falling from the pinnacle of wealth to poverty is rare.
    d. education is the best predictor of intergenerational mobility.

True/False Questions

T F 1. Most sociologists use the concept of the caste system of stratification in reference to India.

T F 2. South Africa is the only country on the African continent working to establish democracy.

T F 3. In South Africa, there are clear physical differences between people classified as Coloured and White.

T F 4. People classified as Coloured, Asian, and African in South Africa are regarded by whites as "blacks."

T F 5. In a true class system, ascribed characteristics do not determine social class.

T F 6. In a true class system, there is no inequality.

T F 7. Functionalists Davis and Moore argue that it is difficult to document the functional importance of an occupation.

T F 8. Karl Marx maintained in all of his important writings that every society contains two social classes.

T F 9. Two of every three poverty-level households are headed by women.

T F 10. Distinct structural factors have contributed to the existence of the urban and the rural underclass in the United States.

Continuing Education

Leisure

For a change in music, listen to

Fosatu Worker Choirs (South African Trade Union Choirs). Rounder.
Rhythm of Resistance: Music of Black South Africa. Earthworks/Virgin.
Sounds of Soweto: The Essential African Music Album. Capitol.
Cruel, Crazy, Beautiful World (Johnny Clegg and Savuka). Capitol.
Third World Child (Johnny Clegg and Savuka). Capitol.

The next time you rent a movie, consider.

A Dry White Season, directed by Euzhan Palcy, a black South African woman (105 minutes). In this powerful film a white Afrikaner confronts the reality of apartheid after his gardener is murdered by the South African police. As you view the film, consider the various ideologies that whites employed to justify the systematic discrimination against the nonwhite people of South Africa.

Travel

If you're in Chicago, visit The Peace Museum (430 W. Erie Street; 312-440-1860). This museum contains exhibits related to peacemaking, peacekeeping, human rights, antiwar protests, and social justice. Artifacts on display include popular music reflecting these peace and social justice themes, antiwar posters, and folk art.

General and Readable Information

Africa News (P.O. Box 6684, Syracuse, NY 13217-7917) is a bi-weekly publication that covers news about every country on the African continent. *Africa News* is a service agency supplying news and feature material to broadcast and print media.

BBC Focus on Africa (Bush House, P.O. Box 76-Strand London WC2b 4PH, U.K.) is a quarterly pubication produced by the BBC African Service and its listeners. Each issue includes news features and updates chronically important events which have occurred since the last issue. It also publishes short stories, poetry, letters to the editor, and general interest articles.

Country Profile
South Africa

Official Name: Republic of South Africa
Population (1994 est.): 47,000,000
Land Area: 471,440 square miles
Population Density: 87 people per square mile
Birth Rate per 1,000: 34
Death Rate per 1,000: 8
Rate of Natural Increase: 2.6%
Average Number of Children Born to a Woman During Her Lifetime: 4.5
Infant Mortality Rate: 52 per 1,000 live births
Life Expectancy at Birth: 64
Gross National Product per Capita: $2,520
Capital: Pretoria

Source: Haub and Yanagishita (1994).

In spite of having the most vibrant economy on the continent, with a Gross Domestic Product equal to 75 percent of that of the rest of sub-Saharan Africa, [the] new government is inheriting a catalogue of problems that will be enough to test the best economists.

A recession in the first three years of this decade further weakened an economy already ravaged by the sanctions campaign, high inflation and inefficient public administration. To this must be added the unrelenting violence which has discouraged investment by both South Africans and foreign businesses in certain areas of the country.

But perhaps the greatest single challenge to be faced is the staggering level of unemployment, estimated by the South African Reserve Bank to be around 46 percent. Even in the mining industry, once the back bone and still responsible for 28 percent of the country's foreign exchange earnings, the numbers employed have fallen by 23 percent since 1988 to 355,000.

Most of those facing unemployment are black. Poorly educated, with only 38 percent of black pupils having passed their school matriculation last year compared with 90 percent of whites, many have had little choice but to opt for the informal sector. Many have become hawkers and their organization (ACHIB) estimates they are now part of the fabric of city life in South Africa.

But the lack of job opportunities, compounded by the naked discrimination of apartheid, is not limited to those at the bottom end of the scale. The room at the Carlton Hotel may have been full but black organizations estimate that fewer than five percent of senior managers in South Africa are black and less than one percent of blacks occupy executive positions.

Affirmative action as a means of addressing black exclusion from jobs and among the senior ranks of management has become one of the hottest topics of discussion in South Africa today. Many companies and public corporations have already started implementing affirmative action programmes.

But Joe Ndelele, a black senior manager at the huge state transport organisation Transnet, is sceptical: "We shall see many organisations pretending they are introducing affirmative action programmes but when you look closely, one must wonder how serious they are including my own organisation. You will find there is a lot of window dressing just to pretend they are doing something."

A manager at another organisation who asked to remain anonymous said, "I think legislation is the answer. I cannot see white people giving up jobs and power in the corporations without a force of the law."

The African National Congress (ANC) has so far not yet committed itself to legislation on affirmative action. Critics, including firm supporters of the ANC, suggest this is because there is still a fear of alienating the white workforce.

With power now within its sights, the ANC has instead been concentrating on calming nerves about its five year reconstruction and development programme launched in January which is designed to start correcting some of the wrongs of apartheid. The plan provides for the construction of a million new houses, electrification of 2.5 million homes, provision of clean water and sanitation for some of the poorest in the country; redistribution of land, a factor of supreme importance in a country where the white 14 percent of the population owns 90 percent of the land; low priced health care and telecommunications.

The ANC says it has no plans to raise taxes to finance the programme, arguing it would instead be paid for by a refocussing of current priorities and greater efficiency in the use of resources.

Although a far cry from the days when the ANC wanted a state controlled economy, economic analysts have not only expressed doubts about the financial viability of some of its programmes but have also suggested the likelihood of increased government expenditure which could fuel inflation.

But Tito Mboweni, deputy head of the ANC's economic department disagrees: "We have looked at things carefully and it is not our wish to have an inflationary programme."

Delivering growth and benefits to the mass of its people, while at the same time avoiding the pitfalls of increased inflation and huge external borrowing, is a tightrope from which many African governments have fallen. South Africa, though cannot be compared to many African countries at the time of their independence.

It has the best infrastructure in Africa, a first class banking system and insurance services, a stock exchange, abundant natural resources and a pool of skilled labour. The economy is expected to show growth of between one and two percent this year and inflation has fallen to around nine percent.

All of this should act as a magnet for entry and re-entry of foreign companies into South Africa after an election, and the country may well act as a gateway for many firms to the rest of the continent.

But Sam Malope, a black businessman who owns four bakeries in Boputhatswana, already has plans to conquer the rest of Africa. "I am not waiting for big foreign companies to come into my market. With apartheid gone, I am now making plans to expand all over South Africa, before going into the neighbouring countries. Within a few years I hope to have bakeries as far away as Cairo. I can't see what is to stop me." The number of black award winners in the world of business seems set to increase as the new South Africa dawns.

Source: From "Slicing the Cake," by Joel Kibazo. *BBC Focus on Africa*. 5(2):18-19.

Chapter References

Brislin, Richard W., Kenneth Cushner, Craig Cherrie, and Mahealani Yong. 1990. *Intercultural Interactions: A Practical Guide*. Newbury Park, CA:Sage.

Coles, Robert. 1978. *Privileged Ones: The Well-Off and the Rich in America*. Boston: Little, Brown.

Halberstam, David. 1987. *The Best and the Brightest*. New York: Penguin.

______.1993. *The Fifties*. New York: Villard.

Haub, Carl and Machiko Yanagishita. 1994. *1994 World Population Data Sheet*. Washington, DC: Population Reference Bureau.

Mouer, Ross E. and Yoshio Sugimoto. 1990. *Images of Japanese Society: A Study in the Social Construction of Reality*. New York: Routledge, Chapman & Hall.

Passell, Peter. 1994. "Economic Scene." *The New York Times* (January 27):C2.

Answers

Concept Application

1. Status value; Life chances; Ascribed characteristics
2. Upward mobility
3. Status group
4. Social stratification
5. Life chances

Multiple-Choice

1.c 2.d 3.a 4.d 5.b 6.c 7.d 8.c 9.a 10.b 11.b 12.d 13.b 14.a 15.c 16.b 17.a 18.b 19.a 20.d 21.d 22.c 23.b 24.a 25.b 26.c

True/False

1.F 2.F 3.F 4.T 5.T 6.F 7.T 8.F 9.T 10.T

Chapter 10
Race and Ethnicity

Study Questions

1. What do Arthur Ashe and Ender Bsaran have in common?

2. Who are Afro-Germans?

3. How do the media typically portray international labor migrants? Is this portrayal accurate? Explain.

4. Why pay special attention to Germany in a chapter on race and ethnicity?

5. How did West Germany come to be a major labor-importing country?

6. How do the essay "East Meets West" and the case of German Turks speak to the concept of ethnicity?

7. What are minority groups? What are the essential characteristics of all minority groups?

8. Distinguish between absorption assimilation and melting pot assimilation.

9. What are racist ideologies? Give at least two examples showing how racist ideologies attempt to justify one group's domination over another.

10. What is a stereotype? How are stereotypes perpetuated and reinforced?

11. According to Robert K. Merton, what is the relationship between prejudice and discrimination?

12. Distinguish between individual discrimination and institutional discrimination. Give examples.

13. What is a stigma? How is this concept relevant to issues of race and ethnicity?

14. What are mixed contacts? How do stigmas dominate the course of interaction between the stigmatized and normals?

15. How do the stigmatized respond to people who treat them as members of a category?

16. What explanation can we give for media and politicians' relentless focus on dark-skinned and other highly visible immigrants?

17. Consider that by social convention the people we call Native American are classified as belonging to the Mongoloid race and the people we call Africans are classified as belonging to the Negroid race. In light of the material presented by sociologist Prince Brown, Jr., does such a classification scheme make sense? Why or why not?

Concept Application

Below are five scenarios and the sources from which they were drawn. Decide which concept or concepts covered in Chapter 10 are represented best by each scenario, and explain why. The following concepts are considered:

Assimilation
Discrimination
Dominant group
Ethnicity
Hate crimes
Ideology

Individual discrimination
Institutionalized discrimination
Involuntary minorities
Melting pot assimilation
Minority groups
Mixed contacts
Normals
Prejudice
Prejudiced discriminators
Prejudiced nondiscriminators
Racism
Refugee
Selective perception
Social identity
Stereotypes
Unprejudiced discriminators
Voluntary minorities

Scenario 1

"In the book, an American Asian woman finds her White beau attractive because he is from Connecticut, *not* Canton. He is tall and lanky; he does not have skinny arms like her brothers and father. He is commanding and gets what he wants. Asian men, however, are not depicted as commanding but as arrogant and chauvinistic. My Asian father has never treated my mother arrogantly. He is not short or uncommunicative, either. My father is tall with broad shoulders, a physical attribute inherited by both my brother and me. We have his strong jawline, too. And I have a dimple on my chin like actor Kirk Douglas. An American Asian woman acquaintance made a comment that my brother and I were unlike 'typical' Asian men because we are tall and muscular. Her own brother is tall and muscular! It gets worse. Two strangers from Latin America, on two separate occasions, asked me if I was 'mixed.' Both refused to believe that I was 100% Asian because I did not fit their stereotype of what an Asian should look like. One even referred to my 'big' eyes" (Wang 1994, p. 20).

Scenario 2

"In my neighborhood, we assumed that white people thought they were better than colored people, that they were always on time, that they nearly always knew what they were doing, that they couldn't dance, that they were sloppy at home, that they couldn't be trusted" (Porter 1994, p. 64).

Scenario 3

"The author's theoretical framework is that being Armenian was once linked to relatively homogeneous community characteristics (e.g., Armenian was the language of the family, the church was an ideological and cultural foundation, there were high rates of marital endogamy), yet after three or four generations in this country, the community structures have changed. 'Armenianness' is still there, but primarily as a feeling, a sort of symbolic and emotional template and commitment which are linked to heritage, ancestry, surname, cuisine, and a shared history. How this change occurs--a change from living in a community to a feeling community--and the kinds of constructions which hold individuals as Armenians are in part a psychological expression of culture and history " (Yenogoyan 1994, p. 237).

Scenario 4

"...[I]ntegration can be seen as a two-way process in which the dominant and subordinate sectors interact to forge a new entity, in much the same way as different paints in a bucket. Under integration, the best elements of both the majority and minority culture are merged into a single and coherent national framework across a range of practices, including intermarriage and education" (Fleras and Elliott 1992, p. 62).

Scenario 5

"On occasion, a person's identity is derived from membership at birth and sustained through involvement in groups and communities whose cultural values and social pressures continue to exert a pervasive influence in shaping members' lives. Personal involvement at this level imposes constraints, demands, and responsibilities, as is the case among the Old Order Mennonites and Hutterites" (Fleras and Elliott 1992, p. 50).

Applied Research

Find out the various ways in which illegal and legal immigrants enter the United States each year. Identify the top five countries of origin listed for each category of legal and illegal immigrant. Write a short essay summarizing any noticeable connections between country and immigrant status. Data on this subject are available through U.S. Immigration and Naturalization Service Publications.

Practice Test

Multiple-Choice Questions

1. International labor migrants generally tend to be
   a. unskilled workers.
   b. very diverse.
   c. refugees.
   d. homogeneous.

2. Germany's demand for foreign labor is unlikely to decrease in the future because
   a. the unification of East and West Germany has increased demand for cheap labor.
   b. the demand for agricultural products has increased substantially as a result of the global economy.

c. Germany has one of the lowest birth rates in the world and an aging population.
d. under German law, people can migrate to the country if they can prove German ancestry.

3. The Soviet Communists erected the Berlin Wall
   a. to stop the flow of people from west to east.
   b. in 1945, the year World War II ended.
   c. to prevent the West German government from recruiting workers from Yugoslavia and Greece.
   d. to stop the flow of people from east to west.

4. _________ has profoundly affected the number of people seeking asylum in Germany.
   a. The unification of East and West Germany
   b. The military coup in Haiti
   c. Political change in the former Soviet Union and eastern Europe
   d. The formation of the European Union

5. The most numerous, poorest, and most visibly different foreigners in Germany are
   a. the Turks.
   b. the Italians.
   c. Afro-Germans.
   d. Romanian gypsies.

6. When sociologists Raymond Breton and his colleagues studied ethnicity in Toronto, they asked respondents ________ questions to determine ethnicity.
   a. 20
   b. at least 50
   c. approximately 100
   d. more than 150

7. Theoretically a person's race is based on ________; a person's ethnicity is based on an almost countless number of traits including, for example, ___________.
   a. biological characteristics; shared tradition
   b. historical characteristics; language
   c. cultural traits; family name
   d. a sense of distinctiveness; physical characteristics

8. The separation of East and West Germany for 40 years can be compared to a social experiment in which
   a. twins are presented with different mazes and asked to find their way out.
   b. one of the twins receives the best education and the other receives no education.
   c. two people living in the same house must divide the household.
   d. twins who share a criminal past are separated forcibly and sent to two different boarding schools.

9. Hispanics are those individuals who were born, or declared ancestors were born, in
   a. Spain or in the Latin American countries.
   b. South American countries.
   c. Mexico.
   d. Spain and Portugal.

10. The most controversial quality identified as a characteristic of a minority group is that
    a. people who belong to such a group are treated as members of a category.
    b. minority status is not necessarily based on numbers.
    c. a minority group may be the majority of the population.
    d. membership is involuntary; if people are free to leave the group, they do not constitute a minority.

11. In ___________ assimilation, members of a minority ethnic or racial group adapt to the ways of the dominant group.
    a. melting pot
    b. absorption
    c. involuntary
    d. voluntary

12. According to sociologist Milton M. Gordon, _________ is likely to occur first.
    a. civic assimilation
    b. marital assimilation
    c. structural assimilation
    d. acculturation

13. ____________ is the most important form of assimilation; yet it is the most difficult to achieve.
    a. Civic assimilation
    b. Marital assimilation

c. Structural assimilation
d. Acculturation

14. A clear example of a voluntary minority is
    a. Native Americans.
    b. African Americans.
    c. Cuban Americans.
    d. Mexican Americans.

15. _________ assimilation is a blend that produces a new cultural system.
    a. Melting pot
    b. Absorption
    c. Involuntary
    d. Voluntary

16. Officially Germany is
    a. a melting pot culture.
    b. a nonimmigration culture.
    c. a haven for refugees.
    d. a labor-exporting country.

17. Sociologist Larry T. Reynolds observes that race as a concept for classifying humans is a product of
    a. the 1960s.
    b. the 1800s.
    c. the 1700s.
    d. the 1400s.

18. In *The Mismeasure of Man*, Stephen Jay Gould discussed how
    a. census takers undercount minority populations in the United States.
    b. there are no accurate operational definitions to measure racist behavior.
    c. immigration policies discriminate on the basis of race.
    d. sloppy science has been used to legitimate racism and sexism.

19. When prejudiced people notice only those behaviors which support stereotypes about a particular minority group, they are engaging in
    a. discrimination.
    b. selective perception.

c. a hate crime.
d. visualization.

20. In *Immigrant Workers and Class Structure in Western Europe*, Stephen Castles and Godula Kosack argue that the roots of racist ideologies and prejudice are based on
a. real dislike for the physical appearance of the groups concerned.
b. a need to justify the subordinate position of racial and ethnic minorities in society.
c. years of actual experience with many people from a particular minority group.
d. deficiencies caused by physical characteristics.

21. __________ are likely to initiate hate crimes.
a. Fair-weather liberals
b. All-weather liberals
c. Timid bigots
d. Active bigots

22. Laws in Germany that limit the proportion of foreigners living in an area to 9 percent are examples of
a. institutionalized discrimination.
b. individual discrimination.
c. acculturation.
d. institutionalized assimilation.

23. The 1831 Act Prohibiting the Teaching of Slaves to Read is an example of
a. covert racism.
b. overt institutionalized discrimination.
c. covert institutionalized discrimination.
d. overt institutionalized assimilation.

24. The dynamics underlying stigma are illustrated most clearly by which one of the following examples?
a. Most Germans view Afro-Germans simply as people with dark skin and are surprised to learn that they speak the German language.
b. Most Germans acknowledge that the head scarf worn by some Turkish women signifies region of origin and financial means.
c. According to German law, ancestry determines citizenship.
d. Most Germans do not discriminate against Turks and other dark-skinned people.

25. Goffman used the term "normal" to mean
    a. healthy.
    b. well-adjusted.
    c. majority.
    d. free of pathology.

26. When studying stigmas, Goffman maintained that one should focus on
    a. interaction between the stigmatized and the normals.
    b. the attribute that is defined as the stigma.
    c. the three types of stigmas.
    d. normals.

27. Which one of the following is not a characteristic of mixed contacts?
    a. Sometimes the stigmatized and the normals avoid each other.
    b. The stigmatized are sure that everyone they meet will view them in a negative light.
    c. Normals often view accomplishments by the stigmatized as signs of remarkable and noteworthy capacities.
    d. Stigmatized persons experience invasions of privacy.

28. Turks who change their religion from Islam to Christianity, who Germanize their names, or who give up wearing head scarves in order to fit into German society represent
    a. direct attempts to correct a stigma.
    b. indirect responses to their stigma.
    c. melting pot assimilation.
    d. responses aimed at changing the way normals view them.

29. Women of ________ ancestry were the first mothers of children born to European and African men in the earliest phases of U.S. history.
    a. European
    b. African
    c. Native American
    d. Spanish

30. Native Americans and Africans share a number of social traits and values that facilitated their absorption into each other's culture. Which one of the following is not one of those traits?
    a. Recognition of kinship traced patrilineally

b. Polygyny
c. A spirituality that celebrated the mutual dependency of humans and animals
d. A reverence for the earth

True/False Questions

T F 1. Afro-Germans do not qualify for German citizenship.

T F 2. Officials from labor-exporting countries supported the West German government's efforts to recruit workers from their countries after World War II.

T F 3. When employment rates are high, the United States stops recruiting international labor migrants.

T F 4. In the United States, Hispanics possess a firm collective self-identity.

T F 5. Discrimination is a behavior, not an attitude.

T F 6. Sociologists Jack Levin and Jack McDevitt found that almost no hate crimes involve blacks attacking whites.

T F 7. Sociologists have conducted considerable research on racist thinking among blacks.

T F 8. Until the fall of the Berlin Wall in 1989 and the subsequent reunification of Germany, East Germans had little or no interaction with non-German people.

T F 9. A person's ethnicity is based on biological traits.

T F 10. The material presented in the textbooks suggests that many Turks in Germany are more German than people of German heritage who did not grow up in Germany.

T F 11. It is impossible to identify race simply through physical features.

Continuing Education

Leisure

For a change in music, listen to

German Favorite Songs. (Ernst Wolff) Smithsonian/Folkways.

The next time you rent a movie, consider

Journey of Hope, a film written and directed by Xavier Koller (111 minutes) and winner of the 1990 Academy Award for best foreign film. The film is based on the true story of a Kurdish couple living with their seven children in a village in Turkey. Haydar, the husband, decides to sell his possessions and emigrate illegally to Switzerland, which he believes is a "paradise." Haydar convinces his wife, Meryem, to accompany him, and at his father's urging, decides to take one of their sons. As you view the movie, consider the various viewpoints and reactions of people whom the couple encounter as they embark on their journey in search of a new life.

Travel

If you're in New York City, visit Ellis Island (212-269-5755), which reopened in 1990 after six years of renovation. It now houses a museum documenting 400 years of emigration to the United States, and is a mute witness to the history of America, the land of immigrants.

General and Readable Information

U.S. Immigrant. (Dennis Carey, Attorney at Law, PO Box 257, Woodland Hills, CA 91365-0257). This magazine is published monthly and is a current review and analysis of immigrant rights. It covers legal, bureaucratic, and recent developments related to immigration. Subscribers are offered free legal consultation with an immigration attorney on a toll-free hotline.

Europe. (2100 M Street, NW, Suite 700, Washington, DC 20037; 1-800-627-7961) This magazine of the European Union covers issues that affect the countries within the European Union and the European Union as an entity.

Country Profile
Germany

Official Name: Federal Republic of Germany
Population (1994 est.): 81.2 million
Land Area: 134,930 square miles
Population Density: 602 people per square mile
Birth Rate per 1,000: 10
Death Rate per 1,000: 11
Rate of Natural Increase: -0.1%
Average Number of Children Born to a Woman During Her Lifetime: 1.3
Infant Mortality Rate: 6 deaths per 1,000 live births
Life Expectancy at Birth: 76 years
Gross National Product per Capita: $23,030
Capital: Berlin

The population of the unified Federal Republic of Germany is primarily German; however there are a substantial number of foreign guest workers and their dependents. An ethnic Danish minority lives in the north, and a small Slavic minority known as the Sorbs lives in eastern Germany. Renowned for their economic productivity, Germans are well-educated. Since the end of World War II, the number of youths entering universities has nearly tripled, and the trade and technical schools in the original 11 states of the FRG are among the world's best.

German culture has produced some of the greatest artists and intellectuals of all time. Composers, artists, writers, scholars, and scientists have always enjoyed prestige in Germany.

With per capita income levels approaching $20,000 in the original 11 states, postwar Germany has become a broadly middle class society. A generous social welfare system provides for universal medical care, unemployment compensation, and other social needs. Modern Germans also are mobile; millions travel abroad.

With unification on October 3, 1990, the FRG has started the major task of bringing the standard of living of Germans in the former GDR up to the levels of western Germany. It appears that this will be a lengthy and difficult process, due to the relative inefficiency of the industrial enterprises in the former GDR, the poor infrastructure in this area, the environmental damage in eastern Germany brought on by years of mismanagement under communist rule, and difficulty in resolving property ownership in the former GDR.

Sources: Haub and Yanagishita (1994); U.S. Department of State (1991).

Chapter References

Fleras, Augie and Jean Leonard Elliott. 1992. *Multiculturalism in Canada*. Scarborough, Ontario:Nelson Canada.

Haub, Carl and Machiko Yanagishita. 1994. *1994 World Population Data Sheet*. Washington, DC: Population Reference Bureau.

Porter, A.P. 1994. "Last Words." *Colors* (May-June):64.

U.S. Department of State. 1991. "Germany." *Background Notes* (#7834) Washington, DC: U.S. Government Printing Office.

Wang, George. 1994. "A Few Good Images." *Interrace* (June/July):20-21.

Yenogoyan, Aram A. 1994. Review of *Armenian-Americans: From Being to Feeling Armenian*. *Contemporary Sociology* 23(2):236-37.

Answers

Concept Application

1. Selective perception; Stigma; Mixed contacts; Stereotypes
2. Stereotypes
3. Ethnicity
4. Melting pot assimilation
5. Ethnicity

Multiple-Choice

1.b 2.c 3.d 4.c 5.a 6.d 7.a 8.d 9.a 10.d 11.b 12.d 13.c 14.c 15.a 16.b 17.c 18.d 19.b 20.b 21.d 22.a 23.b 24.a 25.c 26.a 27.b 28.a 29.c 30.a

True/False

1.F 2.T 3.F 4.F 5.T 6.F 7.F 8.T 9.F 10.T 11.T

Chapter 11
Gender

Study Questions

1. Define gender. Why do sociologists find the concept of "gender" useful?

2. Why is the former Yugoslavia the focus for a chapter on gender?

3. Why is gender not an issue in the former Yugoslavia in the same sense as in the United States?

4. In the broadest sense of the word, what is a feminist?

5. Is male/female a clear-cut biological distinction? Why or why not?

6. Distinguish between sex and gender. How does gender influence male/female "biological" differences?

7. How are gender-schematic decisions and gender polarization related?

8. Explain: "One's sex has a profound effect on life chances."

9. What three assumptions underlie sociologist Randall Collin's theory of sexual stratification? What factors does he name as the source of women's subordinate status in relationship to men?

10. In general, how do economic arrangements shape the character of sexual stratification?

11. Explain why women's access to violence control is particularly an issue in wartime. How might this situation be improved?

12. Give some examples showing how socialization operates to teach people society's gender expectations.

13. How do situational theorists explain social and economic differences between men and women? Give at least two examples.

14. What is sexist ideology? How is it reflected in social institutions?

15. Name five areas of women's lives over which the state may choose to exercise control.

16. In what ways have the media distorted research findings about so-called "date rape"?

17. How do norms about appropriate body language reflect the power differences between males and females? What are the implications of differences in body language for professional women at work?

Concept Application

Below are five scenarios and the sources from which they were drawn. Decide which concept or concepts covered in Chapter 11 are represented best by each scenario, and explain why. The following concepts are considered:

Ethgender
Feeling rules
Femininity (or feminine characteristics)
Feminist
Gender
Gender nonconformists
Gender polarization
Gender schematic decision
Hidden rape

Institutional discrimination
Intersexed
Masculinity (or masculine characteristics)
Primary sex characteristics
Secondary sex characteristics
Sexist ideologies
Sexual property
Social emotions

Scenario 1

"Women were widely excluded from jury service until a few decades ago, many years after it was no longer permissible to exclude blacks as a group, and it was not until 1975 that the Supreme Court ruled that states had to maintain a representative jury pool that included women" (Greenhouse 1994, p. A10).

Scenario 2

"For too long, we have turned away from the rape crisis in these institutions, which now hold 1.3 million men and boys. In most of them, rape is an entrenched tradition, considered by prisoners a legitimate way to 'prove their manhood' and to satisfy sexual needs and the brutal desire for power.

The exact number of sexually assaulted prisoners is unknown, but a conservative estimate, based on extrapolations of two decades of surveys, is that more than 290,000 males are sexually assaulted behind bars every year. By comparison, the Bureau of Justice Statistics estimates that there are 135,000 rapes of women a year nationwide, though many groups believe the number is higher" (Donaldson 1993, p. A11).

Scenario 3

"Female students have tended to score lower than males on standardized math aptitude tests. Still, one Federal Department of Education study shows that when math scores were the same, nearly twice as many males pursued physics as females. Female students with a physics aptitude are often not encouraged or recruited. Sheila Tobias, a political science lecturer, says women often do not possess the characteristics males feel are essential--that is, a certain manner and sense of being consumed by science to the exclusion of all else" (Raffalli 1994, p. 26).

Scenario 4

"Women in professional jobs have workplace issues like the glass ceiling and the mommy track. But now there is one for secretaries: rug-ranking.

'If the secretary's pay is based on her boss's status, not on the content of her job, that's rug-ranking--treating her as a perk like the size of his office or the quality of the carpet on his floor,' said N. Elizabeth Fried, a labor consultant based in Dublin, Ohio. 'Secretaries are the only ones in the corporate world whose pay is directly linked to the boss. Instead of a career path of their own, most secretaries have had a hitch-your-wagon-to-a-star reward system' " (Lewin 1994, p. A1).

Scenario 5

"Over the 11 years that my first novel, *Tales of the City*, was a property bouncing from studio to studio, I found that I could stop a script meeting cold with a single question: 'Will Jon and Michael be allowed to kiss?'

The book chronicles the lives of a group of single people--some straight, some gay--who share an apartment house on Russian Hill in San Francisco. My straight characters never seemed to cause the studio executives a moment's pause, but I was told repeatedly that the gay ones could never be shown on television in romantic situations. Sure, they could be there, but only as wisecracking, asexual adornments to heterosexual friends" (Maupin 1994, p. H-29).

Applied Research

Read the essay "The Strongest Woman in the World," and the introduction to it, "The Political of Muscle," both by Gloria Steinem (1994). Write a three-to-five page paper relating three concepts covered in Chapter 11 to the information Steinem presents in the essay and its introduction.

Practice Test

Multiple-Choice Questions

1. Sociologists define _______ as social distinctions based on culturally conceived and learned ideas about appropriate behavior and appearances for males and for females.
   a. sex
   b. gender
   c. sexuality
   d. primary sex characteristics

2. Sociologists find gender a useful concept because
   a. people of the same sex look and behave in uniform ways.
   b. a society's gender expectations are central to people's lives whether they conform rigidly or resist.
   c. people vary in the extent to which they meet their society's gender expectations.
   d. ideas about appropriate behavior for males and for females do not change.

3. One reason why gender is not an issue in the former Yugoslavia in the same sense as in the United States may be that
   a. women are more advantaged than men there.
   b. there is virtually no feminist voice there.
   c. opportunities for mobility are greater for women than for men there.
   d. there is no gender inequality there.

4. ______ are people who claim no national (ethnic) identification and who identify themselves with the country Yugoslavia.
   a. Nationalist Serbs
   b. Tito supporters
   c. Idealists
   d. Yugoslavs

5. Yugoslavia's prosperity began its decline in 1979, when
   a. Josip Tito, Yugoslavia's president, died.
   b. the first feminist organization was formed.
   c. the world recession forced many guest workers to return home.
   d. Slobodan Milosevic, a fervent nationalist, came to power.

6. ______ is a biological concept, whereas _______ is a social construct.
   a. Gender, sex
   b. Sexuality, gender
   c. Sex, chromosomal sex
   d. Sex, gender

7. _______ physical traits not essential to reproduction.
   a. Primary sex characteristics are
   b. Secondary sex characteristics are
   c. Chromosomal sex characteristics are
   d. Gender is

8. _______ are examples of secondary sex characteristics.
   a. Reproductive organs
   b. Distribution patterns of facial and body hair
   c. Chromosomes
   d. Steroid hormones

9. College students make gender-schematic decisions about possible majors if they ask, even subconsciously,
   a. "Would my parents approve of this major?"
   b. "Will this major lead to a high paying job?"
   c. "Are the professors who teach the classes feminists?"
   d. "What is the 'sex' of this major?"

10. In the United States the average woman outlives her spouse by about nine years. This difference in longevity can be explained by all but which one of the following factors?
    a. Women tend to marry men older than themselves.
    b. Men tend to do the most hazardous jobs.
    c. Women's average life expectancy is longer than men's.
    d. Men seek medical care for physical problems, whereas women seek help for emotional problems.

11. Sociologist Randall Collins maintains that the ideology of ________ is at the heart of sexual stratification.
    a. sexual property
    b. gender polarization
    c. sexism
    d. capitalism

12. Sociologist Randall Collins argues that the extent to which women are subordinate to men depends on
    a. gender polarization and the extent to which decisions are gender-schematic.
    b. women's access to agents of violence control and their position relative to men in the labor market.
    c. the type of economic relationship.
    d. women's position relative to men in the domestic sphere.

13. The case of the Khasi society in India illustrates that
    a. women historically have been viewed and treated as men's sexual property.
    b. differences in physical strength between men and women account for women's subordinate status.
    c. women perform more menial tasks and work longer hours than men.
    d. women also can treat men as sexual property.

For questions 14-16, use one from the following set of responses to identify the economic arrangement associated with each statement.

a. Low-technology tribal societies
b. Fortified households
c. Private households
d. Advanced market economies

14. The honored male is the one who is dominant over others, protects and controls his own property, and conquers others' property.

15. Women can offer men an income and other personal achievements.

16. Women try to act as sexually accessible as possible because they offer men sexual access in exchange for economic security.

17. Socialization theorists argue that socialization processes explain
    a. away seemingly biologically-based male/female differences.
    b. a very small proportion of male/female differences.
    c. an undetermined but significant proportion of male/female differences.
    d. almost 70 percent of male/female differences.

18. When child development specialist Beverly Fagot studied children in play groups at age 12 months and then at age 24 months, she found
    a. significant sex differences between the interaction styles of 12-month-old boys and girls.
    b. no significant sex differences between the interaction styles of 24-month-old boys and girls.
    c. that teachers interacted with toddlers in gender-polarized ways.
    d. that teachers responded to girls when they behaved assertively and to boys when they communicated in gentle ways.

19. Which statement would a situational theorist use to explain gender differences?
    a. A person's position in a social structure can channel behavior in stereotypical male or female directions.
    b. Children's toys figure prominently in the socialization process.
    c. There is a close correspondence between primary sex characteristics and athletic capability.
    d. Gender inequalities have a physical basis.

20. In her study of neonatal intensive care units, Renee Anspach found that nurses and doctors use different criteria to answer the question "How can you tell if an infant is doing well or poorly?" Specifically, she found that in comparison with physicians, nurses tended to draw on __________ to answer that question.
    a. technical or measurable information
    b. immediate perceptual clues obtained during routine examinations

c. interactional clues
d. medical paradigms

21. Sexist ideologies are structured around several notions. Which one of the following is not one of those notions?
  a. People can be classified into two categories: male and female.
  b. There is a close correspondence between a person's sex and other characteristics such as emotional activity.
  c. Primary sex characteristics explain social and other inequalities.
  d. Secondary sex characteristics are closely related to primary sex characteristics.

22. The concept ethgender
  a. considers the additive effects of race and sex.
  b. merges two ascribed statuses into a single category.
  c. acknowledges the combined effects of race, gender, and state.
  d. merges two achieved statuses into a single category.

23. _________ are an example of a state initiative that defines the "proper way" to reproduce offspring.
  a. Laws that encourage women to be the main socializers of their offspring
  b. Policies that govern male soldiers' sexual access to women outside military bases
  c. Policies that limit women's combat roles
  d. Laws prohibiting sexual relationships with persons of another race and ethnicity

24. The term ________ was used originally to capture the differences between the percentage of college women who said they had experienced a rape or an attempted rape and the percentage of women who reported such crimes.
  a. date rape
  b. acquaintance rape
  c. hidden rape
  d. campus rape

25. In their study of male and female college students attending a southern university, Johnson and his colleagues found that ________ of the females miscommunicated their sexual intentions to their male partners.
  a. 66 percent
  b. 50 percent

c. 25 percent
d. 5 percent

26. When sociologists G. David Johnson, Gloria J. Palileo, and Norma B. Gray surveyed more than 1,000 male and female students attending a southern university to replicate the Koss study, they found
    a. an epidemic of date rape on the college campus.
    b. that date rape can be attributed to miscommunication.
    c. considerable miscommunication between men and women.
    d. that the incidence of rape had increased substantially since 1987, when the Koss study was conducted.

27. One reason why women send mixed signals to men about their sexual interests is
    a. so they can cry "rape" if they have "bad sex."
    b. that sending mixed signals is one strategy women use to communicate interests while maintaining their "reputation."
    c. so men will beg them to have sex.
    d. that they are afraid they will not hear again from their male partners.

28. From a sociological viewpoint, the existence of gender nonconformists challenges us to
    a. find ways to "fix" such behavior.
    b. establish clear norms about what constitutes appropriate behavior for males and for females.
    c. critically examine the connections we make between primary sex characteristics and the expression of almost any trait.
    d. identify the biological basis for such behavior.

29. According to Janet Lee Mills in "Body Language Speaks Louder Than Words," professional women have an impossible dilemma. They must simultaneously be
    a. a boss and a mother.
    b. a wife and a boss.
    c. feminine and powerful.
    d. confident and dominant.

True/False Questions

T F 1. Gender was never an issue in the former Yugoslavia in the same way as in the United States.

T F 2. There is no connection between gender and life chances in the former Yugoslavia.

T F 3. No fixed line separates maleness from femaleness.

T F 4. Randall Collins's theory of sexual stratification considers four economic arrangements, one of which is communistic self-management.

T F 5. In the United States, research shows that the husband spends as much time on household tasks as the wife when spouses earn the same income.

T F 6. There is no scientific evidence to support the sexist ideology that homosexuality is incompatible with military service.

T F 7. For the most part, homosexuals and heterosexuals alike are selective in their choice of intimate partners and their expression of sexual behavior.

T F 8. Research evidence suggests that the feminist focus on date rape has created mass hysteria on college campuses.

Continuing Education

Leisure

For a change in music, listen to

Folk Music of Yugoslavia. Smithsonian/Folkways.
Songs and Dances of Yugoslavia. Smithsonian /Folkways.
Yugoslavia National Folk Ballet. Smithsonian/Folkways.

The next time you rent a movie, consider

The documentary *Pumping Iron II: The Women*, directed by George Butler (107 minutes), shows the ways a society's gender expectations constrain behavior and the appearances of various participants in women's bodybuilding. The film offers an insider's view of the world of female bodybuilding. It focuses on the ways judges, the contestants and the audience react to a watershed event in the history of the sport. That event was the presence of a female competitor Bev Francis, a former ballerina, turned power lifter, turned bodybuilder, who had gone way beyond what other female bodybuilders had achieved to that point. As you view the film pay close attention (1) to debates among participants and judges about the meaning of femininity and (2) to how contestants (consciously or unconsciously) focused on preserving their femininity first and developing their muscles second.

Travel

If you're in Seneca Falls, New York, visit the National Women's Hall of Fame (76 Fall Street; 315-568-2936). The Hall of Fame was established in 1969 and is located where the first convention for Women's Rights was held in 1848. There are displays documenting women's accomplishments ranging from Helen Stephens who won two gold medals in the 1936 Olympic Games in Berlin to Dolores Huerta, co-founder of the United Farm Workers with Cesar Chavez. Last year 35 women were inducted into the Hall of Fame.

General and Readable Information

Eastern European Quarterly (1200 University Avenue, Boulder, CO 80309). A scholarly journal in existence for 28 years. It offers in-depth analysis of historical and contemporary issues affecting eastern Europe.

Country Profile
Bosnia and Herzegovina

Official Name: Republic of Bosnia and Herzegovina
Population (1994 est.) : 4.6 million
Land Area:19,740 square miles
Population Density: 233 people per square mile
Birth Rate per 1,000: 14
Death Rate per 1,000: 7
Rate of Natural Increase: 0.7%
Average Number of Children Born to a Woman During Her Lifetime: 1.6
Infant Mortality Rate: 15.2 deaths per 1,000 live births
Life Expectancy at Birth: 72 years
Gross National Product per Capita (1991 est.): $3,200
Capital: Sarajevo

Bosnia and Herzegovina ranked next to Macedonia as the poorest republic in the old Yugoslav federation. Although agriculture has been almost all in private hands, farms have been small and inefficient, and the republic traditionally has been a net importer of food. Industry has been greatly overstaffed, one reflection of the rigidities of Communist central planning and management. Tito had pushed the development of military industries in the republic with the result that Bosnia hosted a large share of Yugoslavia's defense plants. As of March 1993, Bosnia and Herzegovina was being torn apart by the continued bitter interethnic warfare that has caused production to plummet, unemployment and inflation to soar, and human misery to multiply. No reliable economic statistics for 1992 are available, although output clearly fell below the already depressed 1991 level.

Country Profile
Croatia

Official Name: Republic of Croatia
Population (1994 est.): 4.8 million
Land Area: 21,830 square miles
Population Density: 220 people per square mile
Birth Rate per 1,000: 10
Death Rate per 1,000: 11
Rate of Natural Increase: -0.1%
Average Number of Children Born to a Woman During Her Lifetime: 1.4
Infant Mortality Rate: 11 deaths per 1,000 live births
Life Expectancy at Birth: 71 years
Gross National Product per Capita (1991 est.): $5,600
Capital: Zagreb

Before the dissolution of Yugoslavia, the republic of Croatia, after Slovenia, was the most prosperous and industrialized area, with a per capita output roughly comparable to that of Portugal and perhaps one-third above the Yugoslav average. Croatian Serb Nationalists control approximately one third of the Croatian territory, and one of the overriding determinants of Croatia's long-term political and economic prospects will be the resolution of this territorial dispute. Croatia faces monumental problems stemming from: the legacy of longtime Communist mismanagement of the economy; large foreign debt; damage during the fighting to bridges, factories, powerlines, buildings, and houses; the large refugee population, both Croatian and Bosnian; and the disruption of economic ties to Serbia and the other former Yugoslav republics, as well as within its own territory. At the minimum, extensive Western aid and investment, especially in the tourist and oil industries, would seem necessary to salvage a desperate economic situation. However, peace and political stability must come first. As of June 1993, fighting continues among Croats, Serbs, and Muslims, and national boundaries and final political arrangements are still in doubt.

Country Profile
Serbia and Montenegro

Official Name: Yugoslavia
Population (1993 est.): 10.5 million
Land Area: 26,940 square miles
Population Density: 390 people per square mile
Birth Rate per 1,000: 14
Death Rate per 1,000: 10
Rate of Natural increase: 0.4%
Average Number of Children Born to a Woman During Her Lifetime: 1.9
Infant Mortality Rate: 16.5 deaths per 1,000 live births
Life Expectancy at Birth: 72
Gross National Product per Capita (1992 est.): $2,500-$3,500
Capital: Belgrade

The swift collapse of the Yugoslav federation has been followed by bloody ethnic warfare, the destabilization of republic boundaries, and the breakup of important interrepublic trade flows. The situation in Serbia and Montenegro remains fluid in view of the extensive political and military strife. Serbia and Montenegro faces major economic problems. First, like the other former Yugoslav republics, it depended on its sister republics for large amounts of foodstuffs, energy supplies, and manufactures. Wide varieties in climate, mineral resources, and levels of technology among the republics accentuate this interdependence, as did the Communist practice of concentrating much industrial output in a small number of giant plants. The breakup of many of the trade links, the sharp drop in output as industrial plants lost suppliers and markets, and the destruction of physical assets in the fighting all have contributed to the economic difficulties of the republics. One singular factor in the economic situation of Serbia and Montenegro is the continuation in office of a Communist government that is primarily interested in political and military mastery, not economic reform. A further complication is the imposition of economic sanctions by the UN.

Country Profile
Slovenia

Official Name: Republic of Slovenia
Population (1994 est.): 2 million
Land Area: 7,820 square miles
Population Density: 255 people per square mile
Birth Rate per 1,000: 10
Death Rate per 1,000: 10
Rate of Natural Increase: 0.1%
Average Number of Children Born to a Woman During Her Lifetime: 1.3
Infant Mortality Rate: 6.6 deaths per 1,000 live births
Life Expectancy at Birth: 73 years
Gross National Product per Capita (1992 est.): $6,330
Capital: Ljubljana

Slovenia was by far the most prosperous of the former Yugoslav republics, with a per capita income more than twice the Yugoslav average.... Because of its strong ties to Western Europe and the small scale of damage during its fight for independence from Yugoslavia, Slovenia has the brightest prospects among the former Yugoslav republics for economic recovery over the next few years. The dissolution of Yugoslavia, however, has led to severe short-term dislocations in production, employment, and trade ties. For example, overall industrial production fell 10% in 1991; particularly hard hit were the iron and steel, machine-building, chemical, and textile industries. Meanwhile, the continued fighting in other former Yugoslav republics has led to further destruction of long-established trade channels and to an influx of tens of thousands of Croatian and Bosnian refugees. The key program for breaking up and privatizing major industrial firms was established in late 1992. Bright spots for encouraging Western investors are Slovenia's comparatively well-educated work force, and its developed infrastructure, and its Western business attitudes, but instability in Croatia is a deterrent. Slovenia in absolute terms is a small economy, and a little Western investment would go a long way.

Sources: Haub and Yanagishita (1994); U.S. Central Intelligence Agency (1993).

Chapter References

Donaldson, Stephen. 1993. "The Rape Crisis Behind Bars." *The New York Times* (December 29):A11.

Greenhouse, Linda. 1994. "High Court Bars Sex As Standard in Picking Jurors." *The New York Times* (April 20):A1+.

Haub, Carl and Machiko Yanagishita. 1994. *1994 World Population Data Sheet*. Washington, DC: Population Reference Bureau.

Lewin, Tamar. 1994. "As the Boss Goes, So Goes the Secretary: Is It Bias?" *The New York Times* (March 17):A1+.

Maupin, Armistead. 1994. "A Line that Commercial TV Won't Cross." *The New York Times* (January 9):H-29.

Raffalli, Mary. 1994. "Why So Few Women Physicists?" (Educational Life Feature) *The New York Times* (January 9):26.

Steinem, Gloria. 1994. *Moving Beyond Words*. New York: Simon and Schuster.

U.S. Central Intelligence Agency. 1993. *The World Factbook 1993*. Washington, DC: U.S. Government Printing Office.

Answers

Concept Application

1. Institutional discrimination
2. Hidden rape
3. Sexist ideology; Institutionalized discrimination; Gender polarization; Gender-schematic decisions
4. Institutionalized discrimination
5. Feeling rules

Multiple-Choice

1.b 2.b 3.b 4.d 5.c 6.d 7.b 8.b 9.d 10.d 11.a 12.b 13.d 14.b 15.d 16.c 17.c 18.c 19.a 20.c 21.d 22.b 23.d 24.c 25.a 26.c 27.b 28.c 29.c

True/False

1.T 2.F 3.T 4.F 5.F 6.T 7.T 8.F

Chapter 12
Population and Family Life

Study Questions

1. Define family. What do all families have in common?

2. Why are the topics of population and family life placed together in this chapter?

3. Why is Brazil the focus for a chapter on population and family life?

4. Why is it difficult to choose a photograph to represent "the family"?

5. Under what circumstances are official definitions of a family problematic?

6. What kinds of changes made the Industrial Revolution a "revolution" in Western societies?

7. When referring to countries, how is the dichotomy "industrialized--industrializing" misleading?

8. How are labor-intensive poor countries different from mechanized-rich countries?

9. How did the theory of demographic transition originate?

10. Why is Stage 1 of the demographic transition called the stage of "high potential growth"?

11. According to the model of the demographic transition, which factors contributed to a decline in the death rate? To a rise and then an eventual decline in fertility?

12. Do famines occur because there are too many people and not enough food?

13. In what ways is the industrialization in labor-intensive poor countries fundamentally different from the industrialization that occurred in mechanized-rich countries? (In answering the question, consider differences between urbanization and rural-rural migration.)

14. Distinguish between the demographic trap and the demographic gap.

15. What three major intercontinental flows of migration occurred between 1600 and the early part of the twentieth century? How did they affect the population of Brazil?

16. What is a population pyramid? What shapes can it take?

17. Why is Brazil's official mortality rate likely to be inaccurate?

18. Why is it difficult to state how the Industrial Revolution affected family life?

19. How do increases in life expectancy alter the composition of the family?

20. How is the status of children affected by mechanization?

21. How does urbanization affect family life?

22. How did the Industrial Revolution destroy the household-based economy and lead to the breadwinner system?

23. According to Davis, what strains and demographic factors led to the collapse of the breadwinner system?

24. Define the concept "wider families." List two characteristics that distinguish them from the traditional idea of family.

Concept Application

Below are five scenarios and the sources from which they were drawn. Decide which concept or concepts covered in Chapter 12 are represented best by each scenario, and explain why. The following concepts are considered:

Annual per capita consumption of energy
Cohort
Constructive pyramids
Demography
Demographic gap
Demographic trap
Doubling time
Emigration
Expansive pyramids
Family
Immigration
Infant mortality
In-migration
Internal migration
Migration
Mortality crises
Out-migration
Per capita income
Population (study of)
Population pyramid
Positive checks
Pull factors
Push factors
Stationary pyramids
Theory of the demographic transition
Total fertility
Urbanization

Scenario 1

"By 2025, over 1 billion people in Africa and southern Asia will live under conditions of water scarcity. Many North African and Middle Eastern countries are already faced with absolute water scarcity. In Jordan and Israel, over 3,000 people compete for every flow unit of renewable water. By 2025, virtually all North African countries will be faced with high levels of population pressure on their scarce water resources. And, except for Turkey, all of Western Asia will also experience the highest levels of water scarcity " (Falkenmark and Widstrand 1992, p. 20).

Scenario 2

"The reality is of course that, since World War II, tens of millions of people have opted to leave the quiet of the countryside, either "expelled" by drought, disease, or political strife or drawn by dreams broadcast over transistor radios. Some, like the half-million Guatemalan Indians who travel each winter with their wives and families to the Pacific lowlands to pick coffee and cotton or to cut sugarcane, do so in order to survive in their villages during the rest of the year. But for most, migration is a one-way experience, because those who break with their families and communities, their traditional language, clothes, and food, change too much to be able to return" (Riding 1986, p. 8).

Scenario 3

The population pyramid for Denmark looks more like a rectangle than a pyramid. "Each cohort is about the same size as every other one because the birth rate and the death rate have been low and relatively constant for a long time. This means that each age group is about the same size at birth and, since relatively few people die before old age, the cohorts remain close in size until late in life when mortality rates must rise and eat away at the top of the rectangle" (McFalls 1991, pp. 22-23).

Scenario 4

"The villages were as quiet as death.... In one village I remember we had as our guide a tall, middle-aged peasant who had blue eyes and a straw-coloured beard. When he spoke of the famine in all those villages hereabouts he struck his breast and tears came into his eyes. He led us into timbered houses where Russian families were hibernating and waiting for death. In some of them they had no food of any kind. There was one family I saw who left an indelible mark on my mind. The father and mother were lying on the floor when we entered and were almost too weak to rise. Some young children were on a bed above the stove, dying of hunger. A boy of eighteen lay back in a wooden settle against the window sill in a kind of coma. These people had nothing to eat--nothing at all" (Gibbs 1987, p. 494).

Scenario 5

"The theme of this book is the lives and reactions of certain patients in a unique situation--and the implications which these hold out for medicine and science. These patients are among the few survivors of the great sleeping-sickness epidemic of fifty years ago, and their reactions are those brought about by a remarkable new 'awakening' drug (L-Dopa). The lives and responses of these patients, which have no real precedent in the entire history of medicine, are presented in the form of extended case histories or biographies" (Sacks 1989, p. 1).

Applied Research

Go to the government documents section of your library to find U.S. Census data showing age-sex composition by state. Choose a state with a particularly distinctive age-sex composition. Use the data on that state to draw a population pyramid. In addition, write a description explaining why you think the pyramid looks the way its does.

Practice Test

Multiple-Choice Questions

1. No matter what kind of family people are born into, live with, or form later, their lives are shaped by which of the following key episodes?
   a. Birth, death, marriage, old age
   b. Urbanization, migration, fertility, marriage
   c. Occupation, old age, marriage, where family members must travel to find work
   d. Birth, death, the kind of work family members must do, and where family members must travel to find work

2. One important event that has shaped the key episodes of family life on a global scale is
   a. the Industrial Revolution.
   b. the information explosion.
   c. the 1979 world recession.
   d. World War II.

3. ________ is a marriage system involving one husband and multiple wives.
   a. Monogamy
   b. Polygamy
   c. Polyandry
   d. Polgyny

4. Exogamy is an arrangement in which a person marries
   a. on the basis of love.
   b. someone his or her parents have chosen.
   c. outside one's social group.
   d. inside one's social group.

5. Between 1930 and 1950, the official definition of family in the United States reflected
   a. the upheaval of war.
   b. a high divorce rate.
   c. a more rural America.
   d. the phenomenon of urbanization.

6. The changing definition of family suggests that it is useful to think about the family in terms of
   a. specific memberships.
   b. the procreation function.
   c. events which have transformed the character of family life.
   d. the socialization function.

7. Which one of the following statements is least characteristic of a clan-oriented way of life?
   a. Life expectancy is low.
   b. Division of labor is complex.
   c. Children assume adult roles very early.
   d. People follow patterns that have endured for centuries.

8. Brazil is a(n) _______ country.
   a. labor-intensive poor
   b. mechanized-rich
   c. industrialized
   d. First World

9. If the doubling time of a country's population is 24 years, that country would be classified as
   a. mechanized-rich.
   b. labor-intensive poor.

10. *Favelas* are
    a. urban squatter settlements.
    b. uninhabited regions in Brazil.
    c. drought-stricken areas.
    d. rural areas.

11. The simplest but least useful way to express change in birth and death rates is
    a. in absolute terms.
    b. with fractions.
    c. in age-specific terms.
    d. with percentages.

12. 50/1000 is the highest __________ rate possible for any society.
    a. death
    b. fertility
    c. marriage
    d. crude birth

13. Around 1800 the number of people on the planet reached _________ for the first time in history.
    a. 1,000
    b. 100,000
    c. 1,000,000
    d. 1,000,000,000

14. The Black Plague is an example of
    a. a mortality crisis.
    b. a life expectancy crisis.
    c. a tragedy.
    d. a degenerative disease.

15. From the perspective of Thomas Malthus, a famine is
    a. a mortality crisis.
    b. a preventive check.
    c. a plague.
    d. a positive check.

16. Brazilian parents who are poor are more likely to report the birth of a child than a death because
    a. there is no fee for registering births.
    b. they must travel a greater distance to report births than deaths.
    c. births provide access to national social benefits.
    d. it takes one working day for the bureaucracy to process a birth registration and four to five days to process a death.

17. Movement within the boundaries of a single country is known as
    a. emigration.
    b. immigration.
    c. internal migration.
    d. intracontinental migration.

18. Constrictive pyramids indicate that
    a. all age cohorts in a population are roughly the same size.
    b. a population is composed disproportionately of middle-aged and older people.
    c. a population is composed disproportionately of young people.
    d. each age cohort is progressively smaller than the preceding cohort.

For 19-22, decide whether each statement is characteristic of life
    a. before industrialization.
    b. in the mature phase of industrialization.

19. The economy is consumer-oriented.

20. The chance of a woman dying from pregnancy-related causes is 1 in 10,000.

21. Total fertility is approximately 9.

22. The death rate is at least 50 per 1,000 people.

23. With regard to geographic mobility and its effects on family life, sociologists
    a. can make almost no generalizations.
    b. have found that geographic separation hinders interaction.
    c. have found that geographic separation enhances family relationships.
    d. have found that geographic separation is a liberating experience for family members.

24. In most countries, including the United States, most disabled and frail elderly persons are cared for by
    a. daughters or daughters-in-law.
    b. sons or sons-in-law.
    c. nursing home attendants.
    d. private care nurses.

25. In labor-intensive, poor, nonmechanized, extractive economies, total fertility is likely to be as high as
    a. 50/1000.
    b. 9.
    c. 4.
    d. 20/1000.

26. We know that the Industrial Revolution separated the workplace from the home and altered the division of labor between men and women. More specifically,
   a. the woman came to produce most of what her family consumed.
   b. the economic value of women and children increased.
   c. the man became the link between the family and the wider market economy.
   d. the man's role changed from stressful to carefree.

27. According to sociologist Kingsley Davis,
   a. women's entry into the labor market triggered an increase in the divorce rate.
   b. an increase in the divorce rate preceded married women's entry into the labor market by several decades.
   c. the divorce rate has always been high in the United States.
   d. when the divorce rate reaches 50 percent, unemployed married women consider seeking employment to protect themselves in case of divorce.

28. In the focus essay "The Definition of Family Is Expanding," Marciano and Sussman argue that wider families
   a. are made up of nonkin members.
   b. offer no mechanisms to transfer wealth from one generation to the next.
   c. are ultimately unstable unions.
   d. come into existence to meet economic and emotional concerns.

True/False Questions

T F 1. The universal definition of family is two or more persons living together who are related by blood, marriage, or adoption.

T F 2. For all practical purposes, the Amazon region of Brazil is empty of people.

T F 3. The demographic transition is a model of population growth that applies only to the United States.

T F 4. Rubber trappers spend four or five hours a day collecting and processing the latex into rubber.

T F 5. Innovations in contraceptive technology are the most important factor in the decline in fertility that occurred in Western countries around 1880.

T F 6. Every one of Brazil's major regions have been affected by industrialization.

T F 7. The Brazilian government's official definition of family has not changed in the past 50 years.

T F 8. Brazil's official crude death rate is lower than the crude death rate in the United States.

T F 9. The two-income system is a relatively problem-free arrangement.

T F 10. No adequate definition of "the family" exists.

Continuing Education

Leisure

For a change in music, listen to

Central Brazil: Songs and Dances of the Kaiapo Indians. Gallo.
Brazilliance: The Music of Rhythm. Rykodisc.
Brazil Classics. Sire.

The next time you rent a movie, consider

Eat a Bowl of Tea, directed by Wayne Wang (114 minutes). This film reveals how U.S. immigration laws affected the structure of ethnic Chinese family life in the United States before World War II. In addition, the film offers insights about how second generation immigrants respond to expectations and pressures from their parents--and move away. The geographic size of the United States and the opportunities for mobility make this response easier. Wang has produced and directed two other films about Chinese family life in the United States, *Chan is Missing* and *Dim Sum*.

Travel

Watch for a traveling exhibition of photographs taken by the Brazilian-born photographer Sebastiao Salgado. Salgado is drawn to people who live outside power "in a preindustrial society without any of the machinelike sleekness and instant gratification available in industrial societies" (Brenson 1991, p. Bl).

General and Readable Information

Americas (Organization of American States, 1889 F Street NW, Washington, DC 20006). Each issue of this illustrated bimonthly publication contains interesting and informative feature articles on events, history, and daily life in the countries of North, South, and Central America.

Hemisphere. (Latin American and Caribbean Center, FL International University, University Park, Miami, Florida 33199). A magazine of American and Caribbean affairs is published three times each year and is dedicated to providing debate on the problems, initiatives, and achievements of Latin America and the Caribbean.

Country Profile
Brazil

Official Name: Federative Republic of Brazil
Population (1994 est.): 155.3 million
Land Area: 3,265,060 square miles
Population Density: 48 people per square mile
Birth Rate per 1,000: 25
Death Rate per 1,000: 8
Rate of Natural Increase: 1.7%
Average Number of Children Born to a Woman During Her Lifetime: 3.0
Infant Mortality Rate: 66 deaths per 1,000 live births
Life Expectancy at Birth: 67 years
Gross National Product per Capita: $2,770
Capital: Brasilia

With an estimated population of 150 million, Brazil is the most populous country in Latin America and ranks sixth in the world. Most of the people live in the south-central area, which includes the industrial cities of Sao Paulo, Rio de Janeiro, and Belo Horizonte. Urban growth has been rapid; by 1984 the urban sector included more than two-thirds of the total population. Increased urbanization has aided economic development but, at the same time, has created serious social and political problems in the major cities.

Four major groups make up the Brazilian population: indigenous Indians of Tupi and Guarani language stock; the Portuguese, who began colonizing in the 16th century; Africans brought to Brazil as slaves; and various European and Asian immigrant groups that have settled in Brazil since the mid-19th century. The Portuguese often intermarried with the Indians; marriage with slaves was common. Although the basic ethnic stock of Brazil was once Portuguese, subsequent waves of immigration have contributed to a rich ethnic and cultural heritage.

From 1875 until 1960, about 5 million Europeans emigrated to Brazil, settling mainly in four southern states of Sao Paulo, Parana, Santa Catarina, and Rio Grande do Sul. In order of numbers, after the Portuguese, the immigrants have come from Italy, Germany, Spain, Japan, Poland, and the Middle East. The largest Japanese community outside Japan is in Sao Paulo. Despite class distinctions, national identity is strong, and racial friction is a relatively new phenomenon.

Indigenous full-blooded Indians, located mainly in the northern and western border regions and in the upper Amazon Basin, constitute less than 1% of the population. Their numbers are rapidly declining as contact with the outside world and commercial expansion into the interior increase. Brazilian government programs to establish reservations and to provide other forms of assistance have been in effect for years but are increasingly controversial.

Brazil is the only Portuguese-speaking nation in the Americas. About 90% of the population belongs to the Roman Catholic Church, although many Brazilians adhere to Protestantism and spiritualism.

As its geography, population size, and ethnic diversity would imply, Brazil's cultural profile and achievements are extensive, vibrant, and constantly changing. Popular culture predominates, with a thriving popular music industry, relatively active cinema, and a highly developed television empire, producing an enormous number of soap operas (telenovelas) that have found a world market. The visual arts, especially painting, are lively, while literature and theatre, although important, play a less prominent role in this fast-moving media-oriented society.

Traditionally, Brazilian culture has developed around regional subjects, with the country's northeast normally identified with the national themes, both nativist and Afro-Brazilian, while the urban centers of Sao Paulo and Rio de Janeiro have demonstrated a tendency toward a more international, and European oriented expression. With the post-1964 push to a more integrated national culture, these tendencies have diminished somewhat but remain central to understanding the uniqueness of this vast nation.

Sources: Haub and Yanagishita (1994); U.S. Department of State (1990).

Chapter References

Brenson, Michael. 1991. "Images of People Who Live Outside Power." *The New York Times* (April 5):B1.

Falkenmark, Malin and Carl Widstrand. 1992. "Population and Water Resources: A Delicate Balance." *Population Bulletin* 47(3):1-35.

Gibbs, Philip. 1987. "Famine in Russia, October 1921." Pp. 493-95 in *Eyewitness to History*, edited by J. Carey. Cambridge: Harvard University Press.

Haub, Carl and Machiko Yanagishita. 1994. *1994 World Population Data Sheet*. Washington, DC: Population Reference Bureau.

McFalls, Joseph A., Jr. 1991. "Population: A Lively Introduction." Washington, DC: *Population Bulletin* 46(2):1-40.

Riding, Alan. 1986. "Introduction." Pp. 7-9 in *Other Americas*, by Sebastiao Salgado. New York: Pantheon.

Sacks, Oliver. 1983. *Awakenings*. New York: Dutton.

U.S. Department of State. 1990. "Brazil." *Background Notes* (#7756). Washington, DC: U.S. Government Printing Office.

Answers

Concept Application

1. Demographic trap
2. Urbanization; Migration; Push factors; Pull factors; Internal migration
3. Stationary pyramid
4. Positive check (as defined by Malthus)
5. Cohort

Multiple-Choice

1.d 2.a 3.c 4.c 5.c 6.c 7.b 8.a 9.b 10.a 11.a 12.b 13.d 14.a 15.d 16.c 17.c 18.b 19.b 20.b 21.a 22.a 23.a 24.a 25.a 26.c 27.b 28.d

True/False

1.F 2.F 3.F 4.F 5.F 6.T 7.F 8.T 9.F 10.T

Chapter 13
Education

Study Questions

1. Explain the ongoing "crisis" in American education. What does the ongoing nature of this crisis suggest about American schools?

2. Why was the United States chosen as the country to emphasize for the chapter on education?

3. Distinguish between schooling and formal education.

4. What are the two major conceptions of the functions of schooling?

5. Expand on the statement, "Illiteracy is a product of one's environment."

6. In general, what are some of the major cultural and social factors that affect schooling?

7. What factors encouraged parents to comply with compulsory attendance laws when they were first instituted in the United States?

8. What features of early American education survive today? Explain how they are revealed in assignments.

9. Describe the fundamental characteristics of American education. Explain how each is related to problems in American education.

10. Distinguish between the formal and the hidden curriculum.

11. What is "spelling baseball"? What do children learn when they engage in such educational activities?

12. What hidden curriculum is at work when American students fulfill the typical reading assignment?

13. What is tracking? What is the rationale for tracking? Is this rationale supported by research?

14. Explain how the self-fulfilling prophecy can affect students' academic achievements.

15. Discuss some of the problems associated with using multiple-choice and true/false tests.

16. What did James Coleman reveal about American schools? What was the most controversial finding?

17. How were Coleman's findings "used"? What happened when his recommendation to bus students was implemented?

18. According to Coleman, how did the adolescent subculture emerge?

19. What are the major characteristics of the adolescent status system? How does it reflect values of the society? How does it affect education?

20. If Americans study the success of public education in foreign countries, can they borrow those strategies to improve their system of education? Use Japan as an example.

Concept Application

Below are five scenarios and the sources from which they were drawn. Decide which concept or concepts covered in Chapter 13 are represented best by each scenario, and explain why. The following concepts are considered:

Ability grouping
Catechisms
Formal curriculum
Formal education
Hidden curriculum
Illiteracy
Informal education
Schooling
Self-fulfilling prophecy
Social promotion
Status system
Streaming
Tracking

Scenario 1

"Many of the deaf are functional illiterates.... Hans Furth, a psychologist whose work is concerned with the cognition of the deaf,... argues that the congenitally deaf suffer from 'information deprivation.' There are a number of reasons for this. First, they are less exposed to the 'incidental' learning that takes place out of school--for example, to that buzz of conversation that is the background of ordinary life; to television, unless it is captioned, etc. Second, the content of deaf education is meager compared to that of hearing children: so much time is spent teaching deaf children speech--one must envisage between five and eight years of intensive tutoring--that there is little time for transmitting information, culture, complex skills, or anything else.

Yet the desire to have the deaf speak, the insistence that they speak--and from the first, the odd superstitions that have always clustered around the use of sign language, to say nothing of the enormous investment in oral schools allowed this deplorable situation to develop, practically unnoticed except by deaf people, who themselves being unnoticed had little to say in the matter" (Sacks 1989, pp. 28-29).

Scenario 2

"In 1897, Captain Richard Pratt arrived in Sioux country to enlist Sioux children for his Carlisle Indian Industrial School, the first and most famous of what would become a whole system of off-reservation boarding schools for Indian students. Eighty-four Sioux children from Pine Ridge and Rosebud, about two-thirds boys and mainly from prominent families, returned east with the stern captain. Neither parent nor pupil foresaw the short hair, the starched shirts and squeaky boots, the Christian names, or the other trappings....Head shaving and even shackling with a ball and chain were common punishments for Indian pupils who ran away or spoke in their native tongue. Suppressing the Sioux language rated high among both the Indian Bureau's educational priorities and the reasons Sioux parents kept children at home" (Lazarus 1991, pp. 101-03).

Scenario 3

"Given a paycheck and the stub that lists the usual deductions, 26 percent of adult Americans cannot determine if their paycheck is correct. Thirty-six percent, given a W-4 form, cannot enter the right numbers of exemptions in the proper places on the form. Forty-four percent, when given a series of 'help-wanted' ads, cannot match their qualifications to the job requirements. Twenty-two percent cannot address a letter well enough to guarantee that it will reach its destination. Twenty-four percent cannot add their own correct return address to the same envelope. Twenty percent cannot understand an 'equal opportunity' announcement. Over 60 percent, given a series of 'for sale' advertisements for products new and used, cannot calculate the difference between prices for a new and used appliance" (Kozol 1985, p. 9).

Scenario 4

"The development of IQ tests lent an air of objectivity to the placement procedures used to separate children for instruction....Test pioneer Lewis Terman wrote in 1916: 'At every step in the child's progress the school should take account of his vocational possibilities. Preliminary investigations indicate that an IQ below 70 rarely permits anything better than unskilled labor; that the range from 70 to 80 is pre-eminently that of semi-skilled labor, from 80 to 100 that of skilled or ordinary clerical labor, from 100 to 110 or 115 that of the semi-professional pursuits; and that above all these are the grades of intelligence which permit one to enter the professions or the larger fields of business....This information will be a great value in planning the education of a particular child and also in planning the differentiated curriculum here recommended.' " (Oakes 1985, p. 36).

Scenario 5

In the grade schools and high schools of America "various forms of deceit have become accepted as inevitable--passing the incompetent to the next grade to save face; 'graduating' from 'high school' with eighth-grade reading ability; 'equivalence of credits' or photography as good as physics; 'certificates of achievement' for those who fail the 'minimum competency' test" (Barzun 1991, p. 7).

Applied Research

Go to the government documents section of your library and check out the National Endowment for the Humanities (1991) report *National Tests: What Other Countries Expect Their Students to Know*. Read each test question and indicate the questions that you could have answered at the end of your senior year in high school. If you can't obtain this report read "Student Tests in Other Nations Offer U.S. Hints, Study Says," by Susan Chira (1991).

Practice Test

Multiple-Choice Questions

1. The charge "if an unfriendly foreign power had attempted to impose on America the mediocre educational performance that exists today, we might well have viewed it as an act of war" appeared in
   a. *A Nation at Risk*.
   b. *Literacy in the United States*.
   c. "The Education Race: Who Is Ahead of the Class?"
   d. the Coleman Report.

2. The belief that the inadequacies of the school threaten the well-being of the United States and that the schools should be restructured
   a. is unique to the 1980s and 1990s.
   b. emerged when the Soviet Sputnik satellites were launched in the mid-1950s.
   c. has been an issue since the 1960s.
   d. dates back to the founding of mass education in the United States.

3. ________ is a purposeful, planned effort intended to impart specific skills and modes of thought.
   a. Informal education
   b. Formal education
   c. Socialization
   d. Education

4. Most sociological research suggests that schools are most likely to be designed to
   a. liberate minds.
   b. meet the perceived needs of society.
   c. broaden students' horizons.
   d. become aware of conditioning influences around them.

5. One of the most striking characteristics of functional illiterates in the United States today is that they
   a. have had no contact with reading.
   b. do not know the alphabet.
   c. can read and write a little.
   d. cannot do the most basic mathematical problems (i.e., addition).

6. According to the U.S. Department of Education, _______ literacy is the set of knowledge and skills required to apply arithmetic operations.
   a. prose
   b. document
   c. quantitative

7. The fact that Americans perform poorly in academic subjects in comparison with Asians and Europeans suggests that
   a. it is important to learn why this is so.
   b. Americans should emulate European and Asian systems.
   c. European and Asian schools are operated perfectly.
   d. the U.S. system of education has few positive qualities.

8. After thousands of hours of classroom observations and after interviews with both mothers and teachers, Harold Stevenson and his colleagues concluded that American parents are more likely than Taiwanese and Japanese parents
   a. to help their children with homework.
   b. to give high ratings to the quality of education at the schools their children attend.
   c. to encourage their children to do well in school.
   d. to attribute their children's level of achievement to effort.

9. Two of the most prominent features of early American education that have endured to the present are
   a. tracking and dull assignments.
   b. spelling baseball and ambivalence.

c. textbooks modeled after catechisms and single-language instruction.
d. mass education and integration.

10. According to your textbook, true equality in education is realized if everyone
    a. has the opportunity to earn a degree that is valued.
    b. is given the opportunity to attend school.
    c. attends school until they graduate.
    d. is given the opportunity to choose the high school they attend.

11. Which of the following statements about American curriculum is false?
    a. Each of the 50 states sets broad curriculum requirements.
    b. Textbooks, assignments, and instructional methods vary across schools within each state.
    c. Curriculum requirements vary within a school.
    d. There are national guidelines with regard to appropriate curriculum.

12. The sociological significance of the Kentucky Educational and Reform Act (KERA) is that the state recognized that inequality could be corrected only if
    a. funding was tripled.
    b. the way education is delivered to students was completely rethought and overhauled.
    c. everyone had the opportunity to attend college.
    d. teachers were required to take competency tests every five years.

13. In the United States, elementary and secondary schools receive _______ percent of their funding from the federal government.
    a. less than 10
    b. about 30
    c. approximately 50
    d. almost 75

14. French preschool teachers hold the equivalent of a(n) _________ degree in childhood and elementary education.
    a. associate
    b. BS and BA
    c. MA
    d. PhD

15. Most Americans tend to equate education with
   a. increased job opportunities.
   b. personal empowerment.
   c. civic responsibility.
   d. national well-being.

16. The various academic subjects make up the ________ curriculum.
   a. core
   b. hidden
   c. formal
   d. manifest

17. Jules Henry argues that students go along with teachers' requests and participate in activities such as "spelling baseball" because they
   a. don't care whether they learn or not.
   b. are terrified of failure and want badly to succeed.
   c. find such academic "games" entertaining.
   d. find the competition enjoyable.

18. Bruno Bettelheim and Karen Zeland argue that Austrian children learn to read faster and better than American children. They believe that this difference can be explained in part by
   a. teaching methods.
   b. family background.
   c. effort.
   d. the content of the stories.

19. Upon observing three high school classrooms of upper-middle-class, primarily white students in the United States, Israeli social scientist Bracha Alpert found that the students responded
   a. enthusiastically to questions relating to facts and having clear-cut answers.
   b. when teachers asked questions that attempted to stimulate discussions.
   c. in such a muffled way that their responses could rarely be heard by the entire class.
   d. reluctantly to assignments in which they competed in teams to find the right answers.

20. Which of the following is not one of the rationales for tracking?
    a. Students learn better when they are grouped with those who learn at the same rate.
    b. Slow learners develop more positive attitudes when they do not have to compete with the more academically capable.
    c. Students are easier to teach if they have similar academic abilities.
    d. Students in highly academic tracks deserve and have earned the extra time and special attention.

21. In Rosenthal and Jacobson's study of teachers' expectations of students identified as "academic bloomers," the authors found that students identified as "bloomers" improved their test scores over the course of a school year. The researchers concluded that teachers communicated expectation of improvement to "bloomers"
    a. by paying more attention to bloomers than they had in the past.
    b. by giving them extra help before school.
    c. in subtle and complex ways that they could not identify.
    d. through the tone of their voice and extra attention.

22. When A. R. Luria gave illiterate Russian peasants a seemingly simple exercise to which most first-grade pupils are exposed (determining which one of the four objects in a group does not belong),
    a. Luria could not get the peasants to understand the exercise.
    b. Luria found that the peasants could answer the question easily.
    c. the peasants froze and would not answer Luria's questions.
    d. the peasants refused to continue their participation in the study.

23. In comparison to their Asian counterparts, American teachers
    a. work together very closely in preparing lesson plans.
    b. are in the classroom for a smaller portion of the day.
    c. work in environments that discourage systematic learning by students outside the classroom.
    d. receive more systematic training, which prepares them to deal with an array of discipline problems.

24. In *Equality of Educational Opportunity* (the Coleman Report), the single most important variable for explaining differences in test scores among various ethnic groups was
    a. ethnicity.
    b. funding.

c. family background.
d. personal motivation.

25. Brown v. Board of Education (1954) is a famous Supreme Court
 a. desegregation case.
 b. case dealing with the no pass-no play (sports) policy.
 c. school prayer case.
 d. school choice case.

26. Coleman's findings about tests scores and the composition of the student population were used to support
 a. policies designed to encourage residential integration.
 b. white flight.
 c. busing as a means of achieving educational equality.
 d. the Brown v. Board of Education case.

27. According to the Coleman Report, which one of the following factors is least important to academic success?
 a. The social class of one's classmates
 b. Ethnicity
 c. Family background
 d. Peer environment

28. Which one of the following is not a characteristic of the adolescent status system?
 a. For the most part, the peer group is more influential in students' lives than are teachers.
 b. A boy could be the best student and still could be popular if he also dressed well, was a good athlete, and had money for dates.
 c. The female student identified as the brightest has the most friends.
 d. The most admired girls are cheerleaders and those who are successful with the boys.

29. Coleman's study of adolescent subcultures has this important implication for understanding why even the best students in the United States have difficulty in competing with top students in many other countries:
 a. The United States does not draw into competition everyone who has academic potential.
 b. The adolescent subculture's values do not agree with family or classroom values.

c. Parents no longer "train" their children in the skills they know because those skills are outdated and obsolete.
d. Parents exercise more influence than teachers over their children's lives.

30. Students who perceive borders as insurmountable and immerse themselves in the world of their peers would be classified as
a. congruent worlds/smooth transitions.
b. different worlds/border crossings managed.
c. different worlds/border crossings difficult.
d. different worlds/border crossings resisted.

31. Richard Rodriquez's essay *On Becoming a Chicano* speaks to
a. the importance of adolescent subcultures.
b. the nature of formal curriculum.
c. the problems associated with standardized tests.
d. the Americanizing function of education.

32. Although the author of your textbook does not state the connection, we can also say that Rodriquez's essay speaks to
a. the issue of students' multiple worlds.
b. the Americanizing function of education.
c. events leading up to Brown v. Board of Education.
d. the problems associated with standardized tests.

True/False Questions

T F 1. Internationally, American students are conspicuous for the small amount of academic work they do and for how poorly they do.

T F 2. In the strict sense of the word, a person who cannot read a map is illiterate.

T F 3. The typical American student goes to school about 180 days a year. This figure includes field trips, assemblies, and snow days.

T F 4. Sociological research on tracking shows that students learn better when they are grouped with those who learn at the same rate.

T F 5. Sociologist James Coleman maintains that the adolescent society penalizes academic achievement.

T F 6. The United States was the first country in the world to embrace the concept of mass education.

T F 7. The United States has the world's highest post-secondary enrollment ratio.

T F 8. The fact that one college graduate in five is underemployed suggests that level of education is unrelated to income or occupation.

T F 9. Read-do literacy is acquired primarily outside the classroom.

T F 10. Most countries in the world use multiple-choice examinations to measure academic achievement.

T F 11. Coleman's findings about school segregation have changed dramatically over the past three decades.

Continuing Education

Leisure

For a change in music, listen to

Any of the *American Folkways* records distributed by the Smithsonian Institution. For a list of available titles write to Smithsonian/Folkways Recordings, Office of Folklife Programs, 955 L'Enfant Plaza, Suite 2600, Washington, DC 20560.

The next time you rent a movie, consider

Blackboard Jungle, a film directed by Richard Brooks (101 minutes). This movie, released in 1955, is the story of a school teacher trying to connect with students hostile to teachers and to the learning process. The film shows that many of the discipline problems portrayed by the media as new problems were present in the 1950s as well. As you view the film, consider the ways the education problems of the 1950s remain with us today. (Note: The portrayals of male-female relationships reflect the patterns of the 1950s.)

Travel

If you're in Pittsburgh, visit the Nationality Classrooms housed in the 42-story Cathedral of Learning at the University of Pittsburgh (412-624-6000). "These beautiful rooms in the Cathedral of Learning are a very real and important part of education at the University of Pittsburgh. They are classrooms that in themselves are teachers. Their design and decoration are characteristic of many times and many peoples. They represent the best and the noblest heritage of the nationality groups which have helped make Pittsburgh an industrial and cultural capital of the world" (Scaife 1990, p. 2). The rooms were gifts to the university from the various nationality groups of Allegheny County. They include The African Heritage Classroom, The Armenian Classroom, The Chinese Classroom, The Czechoslovak Classroom, The English Classroom, The French Classroom, The German Classroom, The Greek Classroom, The Hungarian Classroom, The Irish Classroom, The Israeli Heritage Classroom, The Italian Classroom, The Lithuanian Classroom, The Norwegian Classroom, The Polish Classroom, The Romanian Classroom, The Russian Classroom, The Scottish Classroom, The Swedish Classroom, The Syria-Lebanon Classroom, The Ukrainian Classroom, The Yugoslav Classroom, and The Early American Classroom.

If you're in Dearborn, Michigan, visit the Henry Ford Museum and Greenfield Village (Village Road at 23 Oakwood Blvd.; 313-271-1620). The museum (12 acres of exhibits) and the 81-acre village (founded by Henry Ford in 1929) are adjacent to one another. The sites were constructed and furnished with the intent of showing the dynamic quality of American history over a 300-year period. The various exhibits demonstrate the extent to which American life was altered by technological innovations (trains, planes, the automobile, and power machinery). The village also includes sites intended to illustrate African-American family life and culture during times of enslavement and freedom.

If you're in Portland, Oregon, visit the American Advertising Museum (9 NW Second Avenue; 503-226-0000). This museum contains more than 200,000 print advertisements, approximately 45,000 billboard advertisements, and a library containing at least 4,000 books about advertising. As you view these exhibits, keep in mind social anthropologist Jules Henry's ideas about the role of advertising in American society. The museum is also recommended as a resource for thinking about values important to American culture.

General and Readable Information

Prospects. (Berman Unipub, 4611-F Assembly Dr., Lanham, MD 20706-4391) This quarterly review of education is a UNESCO publication with an international focus on issues related to education. The journal contains four major sections:
(1) Viewpoints/Controversies, (2) Open File (a series of essays and articles relating to a special topic), (3) Trends/Cases, and (4) Profiles of educators.

Country Profile
United States

Official Name: United States of America
Population (1994 est.): 260.8 million
Land Area: 3,539,230 square miles
Population Density: 74 people per square mile
Birth Rate per 1,000: 16
Death Rate per 1,000: 9
Rate of Natural Increase: 0.7%
Average Number of Children Born to a Woman During Her Lifetime: 2.1
Infant Mortality Rate: 8.3 per 1,000 live births
Life Expectancy at Birth: 76
Gross National Product per Capita: $23,120
Capital: Washington, DC

Source: Haub and Yanagishita (1994)

Typical America contains many racial and ethnic groups. More than 23 million of us (9.3 percent) are of Hispanic origin, and 8 million of us (3.2 percent) identify ourselves as Asians or Pacific Islanders. Blacks number over 31 million, or 12.4 percent of the population, and the American Indian/Eskimo/Aleut populations make up about 2 million (0.8 percent). Nearly 190 million people (75.2 percent) consider themselves to be non-Hispanic white.

According to the 1990 Census, nearly 20 million of us were not born in America but call it home. In 1992, about 33 percent of our population growth came from net immigration. This means America speaks many languages: over 31.8 million of us speak a language other than English at home; over half of this number speaks Spanish.

We are also aging: our median age has risen from 30.0 years in 1980 to 33.1 in 1991 (the median for blacks is 28.1, and the median for whites is 34.1). Nearly 26 percent of us are under 18 years of age, and over 12.6 percent are 65 and older.

Women outnumber men in America by 6.2 million; 51 percent of the total population is female.

There are 95.7 million households in our nation; 70 percent of them contain families. Of all households, 55 percent are maintained by married couples, but only 26 percent of all households with children under 18 include a married couple. About one-fourth of all households contain a person living alone.

About 67 percent of Americans 16 years and over are in the labor force: almost 76 percent of men and nearly 58 percent of women. Over 8 million working-age Americans are prevented from working because of a disability.

Our top three occupation groups are administrative support, including clerical; professional specialty; and executive, administrative, and managerial. There are 19.6 million employed persons working in the retail sales industry.

Of the over 115 million employed workers, 73 percent drive to work alone and 13 percent are in carpools. About 76 percent of us work in the county in which we live, and our commuting time averages about 22 minutes.

Workers are well represented among mothers: almost 58 percent of mothers with children under 6, and 75.9 percent of mothers with children 6 to 17 are in the labor force.

The number of women-owned businesses (sole proprietorship, partnership, or subchapter S companies) increased by a dramatic 57 percent in the late 1980s. Minority-owned firms also dramatically increased in number: Asian-Pacific Islander by 89 percent, Hispanic by 81 percent, American Indian/Eskimo/Aluet by 58 percent, and black by 38 percent.

Most state and local government employees are in education-related activities: two-fifths of the 4.5 million state employees and over one-half of the 10.9 million local government workers.

Personal and business services grew 13 times faster than the population in the late 1980s.

Source: From "United States Population," by Harry A. Scarr. P. 358 in *The World Almanac and Book of Facts 1994*.

Chapter References

Barzun, Jacques. 1991. *Begin Here: The Forgotten Conditions of Teaching and Learning*, edited by M. Philipson. Chicago: University of Chicago Press.

Chira, Susan. 1991. "Student Tests in Other Nations Offer U.S. Hints, Study Says." *The New York Times* (May 20):A1+.

Haub, Carl and Machiko Yanagishita. 1994. *1994 World Population Data Sheet*. Washington, DC: Population Reference Bureau.

Kozol, Jonathan. 1985. *Illiterate America*. Garden City, NY: Anchor.

Lazarus, Edward. 1991. *Black Hills, White Justice: The Sioux Nation versus the United States, 1775 to the Present*. New York: HarperCollins.

National Endowment for the Humanities. 1991. *National Tests: What Other Countries Expect Their Students to Know*. Washington, DC: U.S. Government Printing Office.

Oakes, Jeannie. 1985. *Keeping Track: How Schools Structure Inequality*. New Haven: Yale University Press.

Sacks, Oliver. 1989. *Seeing Voices: A Journey into the World of the Deaf*. Los Angeles: University of California Press.

Scaife, Alan M. 1990. "Foreword." P.2 in *The Nationality Rooms*. Pittsburgh: University of Pittsburgh Press.

Scarr, Harry A. 1993. "United States Population: A Typical American as Seen Through the Eyes of the Census Bureau." Pp. 358-59 in *The World Almanac and Book of Facts 1994*. Mahwah, NJ: Funk&Wagnalls.

Answers

Concept Application

1. Self-fulfilling prophecy
2. Formal education; Schooling
3. Illiteracy
4. Tracking; Ability grouping
5. Social promotion

Multiple-Choice

1.a 2.d 3.b 4.b 5.c 6.c 7.a 8.b 9.c 10.a 11.d 12.b 13.a 14.c 15.a 16.c 17.b 18.d 19.c 20.d 21.c 22.a 23.c 24.c 25.a 26.c 27.b 28.c 29.a 10.d 31.d 32.a

True/False

1.T 2.T 3.T 4.F 5.T 6.T 7.T 8.F 9.T 10.F 11.F

Chapter 14
Religion

Study Questions

1. When sociologists study religion, what do they study?

2. Why was Lebanon chosen as the country to emphasize with regard to religion?

3. How did Lebanon come to possess its geographical boundaries?

4. According to Durkheim, how should sociologists approach the study of religion?

5. According to Durkheim, what are three fundamental and indispensable features of religion? How do these features figure into a definition of religion?

6. Distinguish between the sacred and the profane. What are the three major types of religions, as categorized in terms of sacred phenomena?

7. According to Durkheim, what are rituals? What are the most important outcomes of rituals?

8. Distinguish between ecclesiae, denominations, sects, established sects, and cults.

9. What are some problems with Durkheim's definition of religion? Give examples. Are there better definitions?

10. What functions does religion serve for the individual and the group?

11. Explain what Durkheim means by the statement, "The something out there that people worship is actually society" How is it that society is worthy of such worship?

12. Is religion strictly an integrative force? Why or why not?

13. How did Karl Marx conceptualize religion?

14. What are some criticisms of Marx's conceptualization of religion?

15. According to Weber, what role did the Protestant ethic play in the origins and development of modern capitalism? In what ways has Weber been misinterpreted?

16. What is secularization? Distinguish between Muslim views and American-European views on the causes of secularization.

17. Distinguish between objective and subjective secularizations. To what extent do objective and subjective secularizations exist in American society.

18. What is fundamentalism? How are fundamentalism and secularization related?

19. What are the factors behind the surge of fundamentalism in Muslim countries?

20. How is religion related to the civil war in Lebanon?

21. Use the case of James Dean to discuss why it can be difficult to distinguish between religious and nonreligious activities (e.g., sporting events, graduation ceremonies).

Concept Application

Below are five scenarios and the sources from which they were drawn. Decide which concept or concepts covered in Chapter 14 are represented by each scenario, and explain why. The following concepts are considered:

Church
Civil religion
Classless society
Cults
Denominations
Ecclesiae
Established sects
Fundamentalism
Liberation theology
Modern capitalism
Mystical religions
Objective secularization
Predestination
Profane
Prophetic religions
Religion
Rituals

Sacramental religions
Sacred
Sect
Sectarian community
Secularization
Subjective secularization
This-worldly asceticism

Scenario 1

"As for my own religious practice, I try to live my life pursuing what I call the Bodhisattva ideal....The Bodhisattva ideal is thus the aspiration to practice infinite compassion with infinite wisdom. As a means of helping myself in the quest, I choose to be a Buddhist monk. There are 253 rules of Tibetan monasticism (364 for nuns) and by observing them as closely as I can, I free myself from many of the distractions and worries of life. Some of these rules mainly deal with etiquette, such as the physical distance a monk should walk behind the abbot of his monastery; others are concerned with behaviour. The four root vows concern simple prohibitions: namely that a monk must not kill, steal or lie about his spiritual attainment. He must also be celibate. If he breaks any one of these, he is no longer a monk" (Gyatso 1990, pp. 204-05).

Scenario 2

"The Navajo people don't go out and build churches at every place they regard as holy. The sites that can be of religious importance to them may be utterly indistinguishable [as such to those who think of religion as tied to a church]. A particular bush, a particular tree, a rock, a rise in the landscape [may be regarded as holy]. These sites and the beliefs that are associated with them provide a very basic premise for an entire way of life" (U.S. Commission on Civil Rights 1983, p. 30).

Scenario 3

"The 'miracle' was Brazil's accelerated economic growth between 1968 and 1975; Brazil moved from twenty-first to fourteenth in rank among developing countries, based upon per capita GNP. The 'miracle' did not help most Brazilians, however. The imbalances in the distribution of wealth were made yet worse. The Brazilian bishops have openly denounced the 'Brazilian miracle' for the poverty it has engendered. They have attacked the economic policies that have pushed thousands of peasant farmers off the lands their families have farmed for generations, and they have questioned development projects (such as the exploitation of the Amazon) which displaced the native Indians and poor farmers but brought them no benefit. Indeed, one observer has concluded that 'the church has become the primary institutional focus of dissidence in the country' " (McGuire 1987, p. 215).

Scenario 4

"Mennonites trace their roots to a small group of Christians after 1530 who sought a reformation even more radical than those advocated by Lutherans and Calvinists. They were called Mennonites after Menno Simons, one of their early leaders. Their most distinctive practice is adult baptism offered only to those who have made a decision to follow Christ's teachings" (Lorimer 1989, p. 212).

Scenario 5

"The encyclical, 'Humanae Vitae' ('Of Human Life'), was issued on July 29, 1968--a time of turmoil around the world, a moment of street battles at the Democratic National Convention in Chicago and just between the student uprising in Paris and the Warsaw Pact's invasion of Czechoslovakia.

At conferences observing the encyclical's anniversary, like one attended by 1,400 people, including 14 bishops, in Omaha last week, the defenders are calling it prophetic: they say it accurately predicted that modern societies would slip into promiscuity and exploitation if contraception severed sexual activity from reproduction (Steinfels 1993, p. Y1).

Applied Research

If you are a member of an organized religion, visit a church of another denomination. Choose two sociological concepts from Chapter 14. Write a paper comparing and contrasting the two services or religious events.

Practice Test

Multiple-Choice Questions

1. The sociological study of religion is guided by
   a. strict adherence to the scientific method.
   b. a conviction that there must be one true religion.
   c. the assumption that the supernatural can ultimately be observed.
   d. the belief that some religions are better for a society than others.

2. Lebanon is a microcosm of
   a. the old world order.
   b. the Middle East.
   c. the "new world order."
   d. the Iran-Iraq war.

3. Which one of the following characteristics is most characteristic of a sectarian community?
   a. Geographical diversity
   b. Strong ties to a country
   c. Allegiance to a powerful family
   d. Tolerance for religious diversity

4. At the time of Lebanon's independence, the religious majority in that country was
   a. Maronite Christians.
   b. Sunni Muslims.
   c. Shia Muslims.
   d. Druze.

5. By the early 1980s, the ___________ of Lebanon were the largest single religious community in the country.
   a. Maronite Christians
   b. Sunni Muslims
   c. Shia Muslims
   d. Druze

6. In 1920 the French created the Republic of Lebanon and they drafted a constitution that allocated government power on the basis of
   a. race.
   b. religious affiliation.
   c. political status.
   d. social class.

7. The sociological preoccupation with the definition of religion is rooted in the fact that
   a. it is difficult to prove that a god exists.
   b. it is impossible to make statements about the role of religion in human affairs if we are not clear about the nature of religion.
   c. there are at least eight major world religions.
   d. people use religion to support political aims.

8. The "Enlightened One" is another term for
   a. Buddha.
   b. Confucius.
   c. Jesus Christ.
   d. Abraham.

9. ________ are religions with important geographic and historical ties to one another.
   a. Judaism, Shinto, and Confucianism
   b. Hinduism, Buddhism, and Christianity
   c. Judaism, Islam, and Christianity
   d. Taoism, Judaism, and Hinduism

10. A large number of objects, places, and ideas have been defined as sacred. This fact suggests that
    a. sacred phenomena are "sacred" because of inherent qualities.
    b. there is no relationship between the sacred and the profane.

c. sacredness is a clear-cut quality.
d. ideas about what is sacred are somewhat arbitrary.

11. In mystical religions, the sacred
    a. revolves around items that symbolize historical events.
    b. is sought in states of being.
    c. revolves around the lives, teachings, and writings of great people.
    d. is sought in places, objects, and actions believed to house a god or spirit.

12. Islam is the official religion of
    a. Lebanon.
    b. Bangladesh.
    c. India.
    d. Egypt.

13. Which one of the following is not one of the world's major non-Christian religions?
    a. Buddhism
    b. Confucianism
    c. Taoism
    d. Scientology

14. The Protestant religions owe their origins to ________, who challenged papal authority and many of the practices of the Roman Catholic Church.
    a. Martin Luther King, Jr.
    b. Karl Marx
    c. Emile Durkheim
    d. Martin Luther

15. _________ claim everyone in the society as their members.
    a. Ecclesiae
    b. Denominations
    c. Sects
    d. Established sects

16. _______ definition of religion seems to be the most widely used sociological definition.
    a. Max Weber's
    b. Karl Marx's

c. Emile Durkheim's
d. Robert Coles's

17. In Lebanese society, the _________ commands primary loyalty.
    a. family
    b. neighborhood
    c. religious community
    d. country

Choose the name of the person from the list below that best fits the statement made in 18-24.

a. Karl Marx
b. Emile Durkheim
c. Max Weber

18. The true object of religious worship is society.

19. A belief in the doctrine of predestination created a crisis of meaning among Calvinist adherents.

20. For the individual, religion is the reality from which everything important flows.

21. Religion is the sigh of the oppressed creature, the sentiment of a heartless world, and the soul of soulless conditions.

22. Religion is a source of false consciousness.

23. The Protestant ethic was a significant ideological force in the rise of a capitalist economy.

24. Whenever a group of people has a strong conviction, that conviction almost always takes on a religious character.

25. In 1975, when the civil wars began in Lebanon, two of the poorest segments of the Lebanese population were
    a. the Sunni and the Druze.
    b. the Shia and the Palestinians.

c. the Christian Maronites and the Catholics.
d. the Muslims and the Druze.

26. The belief that __________ placed the greatest pressure on Calvinists to find some sign of salvation.
    a. people are the instruments of divine will
    b. people could change their fate if they worked hard enough
    c. God foreordained all things
    d. not everyone could be saved

27. The two dominant world civilizations at the end of the sixteenth century were
    a. Europe and Africa.
    b. India and China.
    c. Europe and Japan.
    d. the United States and Europe.

28. For the most part, Muslims in the Middle East associate secularization with
    a. an increase in scientific understanding.
    b. modernization.
    c. exposure to the most negative Western values.
    d. fundamentalism.

29. Religious studies professor John L. Esposito prefers the term Islamic __________ to Islamic fundamentalism.
    a. terrorism
    b. militants
    c. revivalism
    d. majority

30. Religion is an important factor in understanding the civil wars in Lebanon. Its explanatory importance lies in
    a. understanding how religion is used by the various factions.
    b. knowing the religious beliefs of the various factions.
    c. understanding the rituals of each faction.
    d. understanding the consequences of fundamentalism.

True/False Questions

T F 1. The various Lebanese civil wars are based on and fought for strictly religious reasons.

T F 2. According to Durkheim's definition of ritual, rituals can be as simple as closing the eyes to pray.

T F 3. Sociologists use the term fundamentalism almost exclusively in reference to two religious groups--the Moral Majority and the Muslims.

T F 4. The Koran was the only book U.S. journalist Terry Anderson read while held hostage in Lebanon.

T F 5. According to Durkheim, the sacred can encompasses evil phenomena.

T F 6. As far as we know, some form of religion has existed as long as humans have lived.

T F 7. The hijab, or modest dress of Muslim women, provides clear evidence that they are severely oppressed.

T F 8. The civil wars in Lebanon can be attributed to the behavior of fanatics acting from primitive and irrational religious convictions.

T F 9. Ironically, the same forces that support secularization processes support the rise of fundamentalism.

Continuing Education

Leisure

For a change in music, listen to

Baal'bek Folk Festival. Monitor.
Days of Fakr Eddeen. Monitor.

The next time you rent a movie, consider

Whistle Down the Wind, a British film directed by Brian Forbes (98 minutes). Three children of strict religious upbringing come across a criminal hiding out in a barn that belongs to their family. They come to believe that he is Jesus Christ, and commit a number of crimes such as stealing food on his behalf. As you view the film, think about its larger message: that people use religion to guide and justify their actions, whether those actions are good or bad.

Travel

If you're in Lancaster, Pennsylvania, visit Pennsylvania Dutch country (contact the Mennonite Tourist Information Center, 2209 Millstream Road; 717-299-0954, or the Pennsylvania Dutch Visitors Bureau; 717-299-8901). Pennsylvania Dutch country contains numerous sites of interest pertaining to the three religious sects that populate the area: Amish, Moravian, and Mennonite. These include Ephrata Cloister (a rural religious community founded in 1732 and disbanded in 1934), and various farms, markets, and museums.

If you're in Savannah, Georgia, visit the Christ Episcopal Church (Johnson Square between E. Julian and E. Congress Streets; 912-232-4131), Congregation Mikve Israel (20 E. Gordon Street; 912-233-1547), the First African Baptist Church (23 Montgomery Street; 912-233-6597), and the Independent Presbyterian Church (Bull Street and W. Oglethorpe; 912-236-3346). All four were founded between 1733 and 1777, although the present buildings were constructed in the mid-to late 1800s.

If you're in Salt Lake City, Utah, visit the Latter-Day Saints Church Office Building (50 E. North Temple Street; 801-240-2331). This 30-story building contains the world's largest genealogical library.

If you're in Santee, San Diego County, California visit the Museum of Creation and Earth History (10946 N. Woodside Avenue; 619-448-0900). "According to museum brochures, 'fossils and other artifacts document the history of the world from its creation to the coming of Christ, and the exhibits, filmstrips, dramas and other displays provide a fascinating and convincing testimony to the fact of creation and the truth of the scriptures' "(Yoachum and Tuller 1993, p. A7).

General and Readable Information

Middle East Insight (1200 18th Street NW, Suite 305, Washington, DC 20036). According to the magazine's advertisement, "Each issue contains lively, stimulating, and timely analyses and opinions from across the spectrum of Middle East perspectives and problems by the best minds in the field. *Middle East Insight* is unique in its mission to promote understanding by providing an impartial forum for the serious presentation of diverse views."

Country Profile
Lebanon

Official Name: Republic of Lebanon
Population (1994 est.): 3.6 million
Land Area: 3,950 square miles
Population Density: 917 people per square mile
Birth Rate per 1,000: 25
Death Rate per 1,000: 5
Rate of Natural Increase: 2.0%
Average Number of Children Born to a Woman During Her Lifetime: 2.9
Infant Mortality Rate: 28 deaths per 1,000 live births
Life Expectancy at Birth: 75 years
Gross National Product per Capita (1991 est.): $1,400
Capital: Beirut

Lebanon has made progress toward rebuilding its political institutions and regaining its national sovereignty since the end of the devastating 16-year civil war in October 1990. Under the Ta'if accord--the blueprint for national reconciliation--the Lebananese have established a more equitable political system, particularly by giving Muslims a greater say in the political process. Since December 1990, the Lebanese have formed three cabinets and conducted the first legislative election in 20 years. Most of the militias have been weakened or disbanded. The Lebanese Armed Forces (LAF) has seized vast quantities of weapons used by the militias during the war and extended central government authority over about one-half of the country. Hizballah, the radical Sh'ia party, is the only significant group that retains most of its weapons. Foreign forces still occupy areas of Lebanon. Israel continues to support a proxy militia, The Army of South Lebanon (ASL), along a narrow stretch of territory contiguous to its border. The ASL's enclave encompasses this self-declared security zone and about 20 kilometers north to the strategic town of Jazzine. As of December 1992, Syria maintained about 30,000 troops in Lebanon. These troops are based mainly in Beirut, North Lebanon, and the Bekaa Valley. Syria's deployment was legitimized by the Arab league early in Lebanon's civil war and in the Ta'if accord. Citing the continued weakness of the LAF, Beirut's requests, and failure of the Lebanese Government to implement all of the constitutional reforms in the Ta'if accord, Damascus has so far refused to withdraw its troops from Beirut.

Sources: Haub and Yanagishita (1994); U.S. Central Intelligence Agency (1993).

Chapter References

Gyatso, Tenzin. 1990. *Freedom in Exile: The Autobiography of the Dalai Lama*. New York: HarperCollins.

Haub, Carl and Machiko Yanagishita. 1994. *1994 World Population Data Sheet*. Washington, DC: Population Reference Bureau.

Lorimer, Lawrence. 1989. "Mennonite Churches." P. 213 in *The Universal Almanac 1990*, edited by J.W. Wright. Kansas City, MO: Universal Press Syndicate.

McGuire, Merridith B. 1987. *Religion: The Social Context*, 2nd ed. Belmont, CA: Wadsworth.

Steinfels, Peter. 1993. "Papal Birth-Control Letter Retains Its Grip." *The New York Times* (August 1):Y1+.

U.S. Central Intelligence Agency. 1993. *The World Factbook 1993*. Washington, DC: U.S. Government Printing Office.

U.S. Commission on Civil Rights. 1983. *Religion in the Constitution: A Delicate Balance* (Clearinghouse Publication No. 80).

Yoachum, Susan and David Tuller. 1993. "Think Tank Tries to Prove Bible Is Literal Truth." *San Francisco Chronicle* (September 14):A7.

Answers

Concept Application

1. Ritual; Mystical religion
2. Sacred; Sacramental religion
3. Liberation theology
4. Sects
5. Fundamentalism

Multiple-Choice

1.a 2.b 3.c 4.a 5.b 6.b 7.b 8.a 9.c 10.d 11.b 12.b 13.d 14.d 15.a 16.c 17.a 18.b 19.c 20.b 21.a 22.a 23.c 24.b 25.b 26.c 27.b 28.c 29.c 30.a

True/False

1.F 2.T 3.F 4.F 5.T 6.T 7.F 8.F 9.T

Chapter 15
Social Change

Study Questions

1. What is social change? Why is it an important topic within the discipline of sociology?

2. How do sociologists study change? What questions do they ask about social change?

3. What are some key events or factors that trigger changes in social life?

4. Why is it difficult to predict the effects of a specific change?

5. Describe the essential dynamics of the Cold War and its legacy.

6. How do sociologists make sense of the seemingly infinite number of interactions and reactions that lead to a specific event?

7. What is an innovation? Distinguish between basic and improving innovations.

8. What is the cultural base? How is the rate of change tied to the size of the cultural base?

9. What is cultural lag? Why did Ogburn emphasize the material component of culture in his theory of cultural lag?

10. Is Ogburn a technological determinist? Why or why not?

11. Ogburn maintains that one of the most urgent challenges facing people today is adapting to material innovations. Does the work of Leslie White lend support to Ogburn's thesis? Why or why not?

12. How does Kuhn define a paradigm?

13. According to Thomas Kuhn, is science simply an evolutionary process? Why or why not? Under what conditions are paradigms threatened? When does a scientific revolution occur?

14. How does Weber define power and authority? Distinguish between legal-rational and charismatic authority.

15. Describe how the charismatic leader is an agent of social change.

16. What are "great changes"? According to C. Wright Mills, who is responsible for such changes?

17. Name the leading institutions in the United States and describe how and when this happened.

18. Does Mills believe that there are any significant constraints on the decision-making power of the power elite? Why or why not?

19. Give an example illustrating how the decisions made by the power elite affect the lives of countless numbers of people on a global scale.

20. How is conflict both a cause and an effect of social change?

21. What are the structural origins of conflict?

22. Name at least three world views or paradigms that emerged in reaction to the Western-style mystique of progress that the economic system of capitalism embodies.

23. From a world system perspective, how has capitalism come to dominate the global network of economic relationships?

24. Distinguish among core, peripheral, and semiperipheral economies.

25. What is the difference between world economy and world-economy?

26. What is the sociological imagination? What is its promise?

27. What factors limit the chances of change in people? Why do people take so long to change? How do people change?

Concept Application

Below are five scenarios and the sources from which they were drawn. Decide which concept or concepts covered in Chapter 15 are represented best by each scenario, and explain why. The following concepts are considered:

Adaptive culture
Anomaly
Authority
Basic innovations
Capitalism
Charismatic authority
Core economies
Cultural base

Cultural lag
Great changes
Improving innovations
Innovation
Legal-rational authority
Paradigms
Peripheral economies
Power elite
Routinized charisma
Scientific revolution
Semiperipheral economies
Simultaneous-independent inventions
Social change
Technological determinism
World-economy

Scenario 1

"Perhaps the best time to have been a Communist in China, I thought, was in the early days of the Party when the idea of making over the country was new, when people could hurl themselves into action, when nothing else mattered, not even death.... Indeed, it was this early spirit that Chairman Mao had hoped to revive in his Cultural Revolution. He himself had thrived on the excitement, the idealism, the feeling of brotherhood of those days, but however electrifying revolutions are at the beginning; they cannot sustain that first passionate fervor. There comes a time when revolutionaries have to return to earth and deal with everyday realities" (Fritz 1985, pp. 77, 79).

Scenario 2

"It is difficult to recapture the medical world of 1800; it was a world of thought structured about assumptions so fundamental that they were only occasionally articulated as such, yet assumptions alien to a twentieth-century medical understanding....The body was seen as a system of intake and outgo, a system that had to remain in balance if the individual were to remain healthy....Equilibrium was synonymous with health, disequilibrium with illness....The physician's most effective weapon was his ability to 'regulate the secretions' to extract blood, to promote the perspiration, the urination, or defecation that attested to his having helped the body regain its customary equilibrium " (Rosenberg 1987, pp. 71-72).

Scenario 3

"Of all the tomorrow makers I would meet, it would be the tomorrows of Eric Drexler that would seem the most outrageously enticing....Consider that you can fit about a million human cells onto the head of a pin. Now consider that Drexler wants to build robotic ships that could, like great paddlewheelers, travel the cytoplasmic sea found within a single cell....Place such a robotic ship into each cell of your body and they might be able to push regenerative DNA keys and repair damaged molecules to keep you forever young. You might be able to change from one person into another. A unified pressing of different DNA keys could result in a change in height, a change in appearance, perhaps even a change in sex....Now if all this seems too much within the realm of science fiction, it is interesting to note that I first heard of Drexler's work while at the U.S. Naval Research Laboratory, when Forrest Carter referred me to a paper of Drexler's that had been published by the National Academy of Sciences" (Fjermedal 1986, pp. 167-68).

Scenario 4

"Thomas Kuhn's seminal work, *The Structure of Scientific Revolutions*, affected working scientists as deeply as it moved those scholars who scrutinize what we do. Before Kuhn, most scientists followed the place-a-stone-in-the-bright-temple-of-knowledge tradition, and would have told you that they hoped, above all, to lay many of the bricks, perhaps even set the keystone, of truth's temple, the additive or meliorist model of scientific progress. Now most scientists of vision hope to foment revolution" (Gould 1987, p. 27).

Scenario 5

In less than three decades, Taiwan has become a major economic player, not only in the economy of the Pacific Rim, but in the global system as well. Foreign investors have played a vital role in Taiwan's economic development. For example, a mass buyer like Sears or K-Mart would visit Taiwanese factories and order goods in bulk for sale under the chain's brand name. A company like Arrow shirts or U.S. Shoe would supply samples to several factories and then contract with the factory that offered the best deal in terms of cost and quality. The "Made in Taiwan" label spread worldwide, even if no one outside Taiwan knew a single Taiwanese company that produced the products (Goldstein 1991).

Applied Research

In Chapter 15 we examined world order during the time of the Cold War, its projected decline, and the simultaneous rise of a "new world order." To learn more about the dynamics of the Cold War, see two films: *Dr. Strangelove or: How I Learned to Stop Worrying and Love the Bomb* (1963) and *Atomic Cafe* (1982). *Dr. Strangelove* pokes fun at the supposedly fail-safe military and government plans devised to ensure that an atomic bomb would not be launched by accident or as a result of the actions of a loose cannon in the ranks of the American government or military. *Atomic Cafe* pieces together footage from various American propaganda and instructive documentaries that were released for public consumption during the Cold War. Taken together, the two movies illustrate the kinds of Cold War issues that preoccupied people in the 1950s and early 1960s and provide insights about the tentative plans that people made in the event of a nuclear war between the United States and the Soviet Union. Conduct a periodical search and find published reviews of these films. Write a short paper about the critics' reactions. Consider how the Cold War politics shown in these films may have contributed to the protests during the 1960s against "authority and the system."

Practice Test

Multiple-Choice Questions

1. When the Strategic Arms Reduction Treaty between the United States and Russia was signed in 1993, each side employed _________ persons to maintain and build their nuclear arsenals.
   a. 100,000
   b. 250,000
   c. 500,000
   d. more than 1 million

2. Identifying the effects of nuclear bomb proliferation is virtually impossible because some materials will remain radioactive for at least ________ years.
   a. 100
   b. 2,500
   c. 25,000
   d. 240,000

3. Which one of the following statements about the Cold War is false?
   a. During the Cold War, relations between the United States and the Soviet Union fell short of direct, full-scale military engagement.
   b. On the Soviet side, the end of the Cold War can be measured by an unprecedented increase in the amount of information for public consumption.
   c. On the American side, the end of the Cold War can be measured by the U.S. government's release of health data that it collected on people who worked at nuclear plants.
   d. The Cold War and its consequences ended with the fall of the Berlin Wall in 1989.

4. Improving innovations are ________ inventions.
   a. modifying
   b. revolutionary
   c. unprecedented
   d. ground-breaking

5. Which one of the following is an example of an improving innovation?
   a. The first nuclear explosion in the New Mexico desert in 1945
   b. The "dry" hydrogen bomb

c. The first sustained nuclear chain reaction in 1942
d. The invention of mass production by Henry Ford in 1904

6. Anthropologist Leslie White maintained that inventions control people. He supported this conclusion with the argument that
   a. necessity is the mother of invention.
   b. when the cultural base is capable of supporting an invention that invention will come into being whether we want it or not.
   c. human beings have no free will.
   d. the best things in science are found because they are useful at the time.

7. Inventors may be geniuses, but they also must be born in the right place and at the right time; this means that
   a. they must be born in a capitalist country free of government control.
   b. the society into which they are born must allow the masses access to education.
   c. they must live in a society with a cultural base sufficiently developed to support their invention.
   d. people must perceive their inventions as useful.

8. After the United States exploded a nuclear bomb, the Soviets, under Stalin, gave absolute priority to
   a. space exploration and arms production.
   b. oil production and mining.
   c. health care and housing.
   d. central planning and social welfare.

9. In the theory of cultural lag, the nonmaterial component of culture is defined as ________ culture.
   a. material
   b. adaptive
   c. latent
   d. manifest

10. Leslie White maintains that if an invention is to come into being, the inventor must
    a. be a genius.
    b. invent something that people view as a necessity.
    c. have the ability to market the product.
    d. be born at the right place and time.

11. People are least likely to accept a new paradigm if
    a. the person presenting the idea has authority.
    b. the idea accommodates important values and traditions.
    c. the idea is forced on them.
    d. it explains an anomaly.

12. An anomaly is
    a. a dominant and widely accepted theory.
    b. a successful application of a paradigm.
    c. an observation that an existing paradigm cannot explain.
    d. a break with an old paradigm.

13. In the sociological sense of the word, a charismatic leader is
    a. a popular person.
    b. demanding to the point of insisting that followers make extraordinary sacrifices.
    c. an attractive, likable, pleasant person.
    d. someone who is continually in our thoughts.

14. The authority of the power elite rests on _______ authority.
    a. legal-rational
    b. charismatic
    c. traditional
    d. political

15. Mills identified the military, the government, and the approximately ________ largest American corporations as the leading institutions in American society.
    a. 10
    b. 50
    c. 200
    d. 500

16. In the United States, ______________________ is an area in which the overlapping interests of the political, the military, and the corporate elite are particularly evident.
    a. advertising
    b. the national interest
    c. environmental protection
    d. defense and arms production

17. Every authority structure contains at least two groups:
    a. the power elite and the masses.
    b. the bourgeoisie and the proletariat.
    c. those with charismatic authority, who lead by virtue of their personality, and those with legal-rational authority, who lead according to the rules.
    d. those with power who have an interest in preserving the system; those without power who have an intent in changing it.

18. Vaclav Havel, the president of the Czech Republic, believes that ________ may have played an important role in causing the revolutions in central Europe.
    a. Chernobyl
    b. the fall of the Berlin Wall
    c. the WIPP project
    d. the collapse of the Soviet Union

19. The information presented on Dr. Alice Stewart, the scientist who managed to shock the world with her reports on atomic workers' health, illustrates
    a. how significant events make powerless people aware that they share an interest in changing the system.
    b. that people will organize, once they realize they have nothing left to lose.
    c. that the U.S. government was open to information which could improve safety in nuclear plants.
    d. that people in positions of authority often use their positions to keep potentially damaging information from those who might use it to change the system.

20. The most recent reaction against the Industrial Revolution's myth of progress is
    a. the Green Party.
    b. super progress.
    c. Islamic fundamentalism.
    d. Christian fundamentalism.

21. People who use the term world economy (without a hyphen) believe that
    a. global interdependence has existed for at least 500 years.
    b. globalization is a relatively new phenomenon.
    c. colonization is at the root of global interdependence.
    d. the global economy involves relationships which transcend national boundaries.

22. Which one of the following is least likely to motivate capitalism?
    a. The pursuit of profit
    b. Competition
    c. Free market
    d. Environmental consciousness

23. Peripheral economies tend
    a. to have strong, stable governments and highly diversified economies.
    b. to have moderately diversified economies.
    c. to have economies dependent on a single commodity.
    d. to exploit weaker economies and to be exploited by stronger economies.

24. Core economies tend
    a. to have strong, stable governments and highly diversified economies.
    b. to have moderately diversified economies.
    c. to have economies dependent on a single commodity.
    d. to exploit weaker economies and to be exploited by stronger economies.

25. "The Case of Nike" is used in the text to illustrate which response by capitalists to create economic growth?
    a. Reducing production costs by introducing labor-saving technologies
    b. Redistributing wealth so as to enable more people to purchase products and services
    c. Expanding the outer boundaries of the world-economy
    d. Improving on existing products and making previous versions obsolete

26. World systems theorists maintain that political upheavals ultimately are caused by
    a. uneven and unequal integration of some groups into the world-economy.
    b. ethnic tensions.
    c. cultural lag.
    d. fundamentalist factions within every society.

27. Understanding the connection between _______ is central to the sociological imagination.
    a. troubles and issues
    b. sociology and the social sciences
    c. material and nonmaterial culture
    d. conflict and social change

28. The time lag between the discovery of the cause of something and the application of the cure is
    a. almost negligible.
    b. about a year.
    c. 10 years.
    d. about one or two generations.

29. The statement "The voice of the intellect is a soft one but it does not rest until it has gained a hearing" was written by
    a. C. Wright Mills.
    b. Sigmund Freud.
    c. Emile Durkheim.
    d. Stephen Jay Gould.

30. Wangari Maathai, who launched the Green Belt movement, mobilized more than 50,000 women to plant more than 10 million trees. These women are known as
    a. Green revolutionists.
    b. Marxist farmers.
    c. foresters without diplomas.
    d. environmental soldiers.

31. The promise of sociology is embodied in
    a. the three theoretical perspectives.
    b. the scientific revolution.
    c. the sociological imagination.
    d. technological determinism.

True/False Questions

T F 1. Karl Marx opposed all aspects of capitalism.

T F 2. The sales of the 10 largest industrial corporations with headquarters in the United States exceed the GNP of Canada.

T F 3. Unlike Robert Oppenheimer, his American counterpart, Andrei Sakharov was not bothered by the moral implications and technical dangers of testing and producing nuclear weapons.

T F 4. People always have choices, even those who seem "trapped."

T F 5. The Worker Adjustment and Retraining Notification Act applies to defense contractors such as General Dynamics Corporation.

T F 6. Over the past 50 years, a handful of people have decided where to locate nuclear production, testing, and waste disposal sites without consulting the local residents or explaining the risks involved.

T F 7. The United States is an example of a country that perfectly embodies the principles of capitalism.

T F 8. The United States government uses the Marshall Islands to test Star Wars technology.

T F 9. William F. Ogburn, who wrote "Cultural Lag as Theory," was a technological determinist.

T F 10. Because change is caused by a seemingly endless sequence of events, sociologists cannot identify key events that trigger change.

Continuing Education

Leisure

For a change in music, listen to

World Music of Struggle: We Shall Overcome. Folkways/Columbia.
Disappearing World (music of endangered cultures from remote regions of the world). Saydisc.

The next time you rent a movie, consider

Dr. Strangelove or: How I Learned to Stop Worrying and Love the Bomb (1963), a film directed by Stanley Kubrick (93 minutes). See "Applied Research" (p. 15.9) for suggestions about how to view this film.

Travel

If you're in Palo Alto, California, visit Stanford University's Hoover Institution (415-723-1754). Currently curators are traveling throughout the Soviet Union and Eastern Europe, collecting political pamphlets, newspapers, posters, and diaries that chronicle the end of the Cold War. The Hoover Institution, founded in 1919, houses one of the world's largest collections of documents pertaining to communism, war, and revolution.

If you're near the White Sands Missile Range, New Mexico, located in the Tularosa Basin between the Sacramento Mountains to the east and the San Andres and Organ Mountains to the west, visit the Trinity Site, a national historic landmark (505-678-1134). According to a United States government pamphlet, "Trinity Site is where the first atomic bomb was tested at 5:29:45 A.M. Mountain War Time on July 16, 1945. The 19-kiloton explosion not only led to a quick end to the war in the Pacific but also ushered the world into the atomic age. All life on earth has been touched by the event which took place here."

If you're in Albuquerque, New Mexico, visit the National Atomic Museum (Kirtland Air Force Base; 505-844-8443). As the name suggests, this museum specializes in the history and effects of nuclear missiles from the first bomb to the A-bomb Minuteman missile.

General and Readable Information

In view of the size of Eastern Europe and the number of republics that made up the former Soviet Union, it is difficult to recommend only one periodical. Three excellent sources of information on the people, events, and issues in this rapidly changing area of the world are *East European Quarterly* (1200 University Avenue, Boulder, CO 80309); *Moscow News* (16/2 Gorky Street, Moscow 9, USSR; subscriptions available through Victor Kamkin, Inc., 12224 Parklawn Drive, Rockville, MD 20852); and *Nationalities Papers* (subscriptions available through Andris Skreija, Department of Sociology, University of Nebraska at Omaha, Omaha, NE 68182).

Chapter References

Fjermedal, Grant. 1986. *The Tomorrow Makers: A Brave New World of Living-Brain Machines*. New York: Macmillan.

Fritz, Jean. 1985. *China Homecoming*. New York: Putnam.

Goldstein, Steven M. 1991. *Minidragons: Fragile Economic Miracles in the Pacific*. New York: Ambrose Video.

Gould, Stephen Jay. 1987. *An Urchin in the Storm: Essays about Books and Ideas*. New York: Norton.

Rosenberg, Charles E. 1987. *The Care of Strangers: The Rise of America's Hospital System*. New York: Basic Books.

Answers

Concept Application

1. Charismatic authority; Routinized charisma
2. Paradigms
3. Innovation; Invention; Discovery
4. Scientific revolution
5. Semiperipheral economy

Multiple-Choice

1.d 2.d 3.d 4.a 5.b 6.b 7.c 8.a 9.b 10.d 11.c 12.c 13.b 14.a 15.c 16.d 17.d 18.a 19.d 20.c 21.b 22.d 23.c 24.a 25.d 26.a 27.a 28.d 29.b 30.c 31.c

True/False

1.F 2.T 3.F 4.T 5.F 6.T 7.F 8.T 9.F 10.F

APPENDIX A

Prepared by Paul Knepper, Ph.D.
Justice Studies Program
Northern Kentucky University

HOW TO FIND SOURCES FOR RESEARCH PAPERS

Finding material for a research paper is easy once you know where to look. On the following pages, how and where to look for sources is presented as a series of questions frequently asked by sociology students

1. I've been assigned a research paper and I don't know where to begin. Where do I look for information about my topic?

Research papers call for library research. If you are not familiar with library services, arrange to take a tour with a librarian. You should plan to do your research at the university library. Local libraries have smaller and less specialized collections, so they probably won't have everything you need. If you have spent some time in the university library before, you already know it's the place you'll need to go. Remember to take paper and a pencil with you in order to write down call numbers, take notes, and so on.

2. I've been to the library, but I can't find anything. Should I change my topic?

If you read about your topic in the textbook or heard about it during a lecture, it's in the library. If you're not sure your topic is researchable because you can't remember how you found out about it, check with your professor. There are library materials for virtually every topic, but library research will proceed quickly and smoothly if you keep three things in mind:

a. Nail down your thesis as soon as possible. If you have only a vague, general idea about what you plan to write about, you will be overwhelmed by the amount of information. As you begin to find materials related to your topic, begin to fashion a thesis in your head. Think about it and write it down. Then refine it and rethink it as you read the sources you've found. Once you know what you intend to say, and how you plan to back it up, you can begin a selective search for appropriate sources.

b. Look for information using key words. Look for materials using the most descriptive words and phrases from your thesis statement. If, for example, you intend to write a paper showing racial discrimination in police use of deadly force, start out by looking under broad categories such as "police," "law enforcement," or possibly "race discrimination." Once you have found listings under these headings, look under more specific categories such as "police use of force" and "deadly force."

c. Remember that each source you find will lead you to others. Check the footnotes, endnotes, and reference lists of every book and article you come across on your topic for additional sources. You're sure to find them, and when you do, they will lead you to still other sources

3. How do I find books on my topic?

The easiest place to find books on your topic is to look through the course textbook. Many textbooks have chapter sections entitled "For Further Reading," or something of the sort, that list books on the same and related topics. Every textbook has a reference list, usually at the back, so be sure to check there as well.

Books on your topic can also be found in the card catalog. It lists books by author, title, and subject. One useful bit of information to know about is that within the Library of Congress classification system, call numbers beginning with "H" are reserved for social sciences.

Sociology books can be found under the call letters HM, HN, HQ, HT, and HV. More specifically, HM 1-299 is reserved for general and theoretical sociology; HN 1-981 is for social problems; HQ 1-2039 is reserved for the fancily; HT 51-1595 is for books on communities, social class, and race; and HV 1-9960 is reserved for social pathology.

Besides the card catalog, books (and other sources) about a particular topic can be found in published bibliographies. Bibliographies are typically arranged by subject, and they supply all the information needed (title, author, year of publication, etc.) to locate the book you want. Annotated bibliographies provide a short summary of each source along with the bibliographic information. Usually, bibliographies are located in the reference section of the library.

4. I've heard the term *scholarly journal.* What is it? Why should I bother to look for one?

A scholarly journal is a periodical published for an academic or a professional audience. Some are published by academic associations, such as *American Sociological Review* (published by the American Sociological Association). Sociological journals typically appear four times a year (quarterly), and each issue contains four or five articles, along with book reviews, research notes, and editorials. At most libraries the individual issues of a journal are bound together for each year and are shelved like books; the title, volume or issue numbers, and year of publication appear on the binding. If they are not on the shelves in bound form, they are likely to be on microfilm or microfiche.

Journal articles are one of the best sources of information because they provide current, credible, and concise information on a wide variety of topics. They are written by researchers and other professionals who display a great deal of expertise. The following is a list (not comprehensive) of some sociological journals:

American Journal of Sociology
American Sociological Review
American Sociologist
British Journal of Sociology
Contemporary Sociology
Journal of Health and Social Behavior
Social Forces
Social Problems
Social Psychology Quarterly
Sociological Analysis
Sociological Focus
Sociological Inquiry
Sociological Practice Review
Sociological Quarterly
Sociological Review
Sociology and Social Research
Sociology of Education
Sociology of Sport Journal
Urban Review

5. I'd like to find some journal articles on my topic. How do I do that?

You can find articles dealing with your topic by using indexes and abstracts. Indexes give the title, author, source, pages, and date of publication of journal articles under specific headings, usually author and topic. Abstracts are like annotated bibliographies-in addition, they give a brief summary of the article. Indexes and abstracts can be found in the reference section of the library. Some of the most useful indexes and abstracts in Sociology are:

Sociological Abstracts
Social Sciences Index

Once you know the name of the journal you want, all you need to know is the call number to find it on the shelf. Ask a librarian to show you the list of periodicals (sometimes called the "serials printout") subscribed to by the library. It lists all the journals the library has, usually in alphabetical order by title, and shows the periodicity, format, holdings, and location of each. *The Annual Review of Sociology* (Volumes 1-18) is another important tool.

6. I'd like to include some statistics in my research paper. Where can I get current facts and figures on my topic?

Government publications are the foremost source of statistics. There are several government agencies, such as the Bureau of the Census, that publish statistics on many areas of life. Some of the most general and accessible sources are:

City and County Databook. (annual) Washington, DC: U.S. Bureau of the Census.
Statistical Abstract of the United States. (annual) Washington, DC: U.S. Bureau of the Census.
World Population Profile. (annual) Washington, DC: U.S. Bureau of the Census

Besides government publications, there are a variety of directories and handbooks. These are typically published by national organizations, and they provide a wealth of facts and figures about various topics. *The Gallup Report* is an example of a private organization that provides valuable information on public opinion. Directories and handbooks can usually be found in the reference section of the library. Other examples are:

Information Please Almanac. (annual) Boston: Houghton Mifflin Company (general statistics on a variety of topics).
Human Development Report. (1991) United Nations (a variety of statistics on life in countries around the world).

7. Are there any sources I shouldn't use for my research paper?

There are several kinds of materials found in a library that you should avoid using as primary sources for a research paper done for a college course. Avoid local newspapers and the most popular magazines, such as *Time, USA Today, Reader's Digest, Newsweek, U.S. News and World Report,* and *Psychology Today* as your primary sources. Likewise, you should not use popular encyclopedias or other home-reference books, such as *World Book Encyclopedia* and *Encyclopaedia Britannica*, as exclusive sources. Although these sources might provide some good background reading, they have become associated with poor research practices acquired during grade school and high school: When the teacher assigned a research paper many of us simply went to the encyclopedia or a popular magazine and copied the material or reworded it. There are other problems as well, especially if the source is a local newspaper or popular magazine:

a. The information is limited. Since the articles in these publications are written for popular consumption, they are not likely to include detailed information about all aspects of a topic nor are they likely to present a wide range of topics and issues. Often, only the most sensational stories make good copy.

b. The analyses are often overly simplistic. Since the expressed goal of the popular publications is profit, editors are not always concerned with contributing to a field of knowledge, nor is their prime goal to make responsible public policy recommendations. When difficult and controversial issues are included, they are typically framed in an unduly simplistic way.

For a change you might want to try sources such as *The New York Times, The Wall Street Journal, The New York Review of Books, Atlantic Magazine, The Washington Post, Daedalus,* and *The American Scholar*. Also make it a point to find and use articles in scholarly journals that relate to your topic. An important rule of thumb is that you should not rely on one or two sources, especially if they are encyclopedias, local papers, or popular magazines.

8. I'd like to get a good grade on my research paper. Is there any source I can use that will really impress my instructor?

Use of scholarly journals will definitely impress your instructor. Journals cover the widest range of topics and present information in relatively brief (compared to books), digestible portions. Journals that are peer-reviewed offer some of the best sources.

The term *peer-reviewed* refers to the process each article must go through before it is published. First the author sends several copies of the manuscript to the editor, who sends copies to a number of reviewers--other researchers who specialize in the same area. Each reviewer independently and objectively considers the merits of the research, then provides written comments for the author and the editor. If the balance of the reviews is favorable, and the editor decides the issue is significant, and the author is willing to address the concerns of reviewers, the article appears in print. This careful, sometimes cumbersome, peer-review process is what makes scholarly journals some of the best possible sources of information for your research paper.

HOW TO WRITE A RESEARCH PAPER

Many students believe that first-rate papers are written by individuals born with exceptional literary skills. Other students claim that it's impossible for them to write superlative papers because they performed poorly in high school English. Still other students are confident that great papers can be written the night before the due date and are surprised to discover that the instructor did not agree with them.

Truth is, any student can write a top-quality research paper provided she or he is willing to spend some time thinking, researching, writing, and revising. Writing a paper for a college course is both easier and harder than most people think. It's easy once you know the basic rules for putting together a research paper. It's hard because it takes time and effort.

What Is a Research Paper?

A research paper for a college course makes a statement about a particular issue through the organized presentation of research material. You make a claim about an important topic and then systematically back up that claim with credible facts, reasoned argument, and considered opinion.

When college professors (or other professional researchers such as those employed by think tanks and government agencies) write research papers, they usually draw conclusions from the findings of research they have conducted themselves. This kind of research paper reports primary-research information gathered firsthand through opinion polls, statistical analyses, interviews, or other methods. When college students write research papers, they draw conclusions on the basis of secondary research. Student papers make use of published original research by professional researchers (in the form of books and articles) available at the library.

In other words, student papers are derived from other people's work. Research papers by college students typically do not present original research but borrow facts and opinions from published sources of information. But at the same time, your research paper should be original in that it represents your organization of the material, your writing, and your conclusions. If you fail to do your own work, it may constitute plagiarism.

Plagiarism means to steal or use the ideas or language of another writer with or without intent to deceive the reader. Plagiarism includes: copying paragraphs, sentences, or significant parts of sentences without using quotation marks; paraphrasing without reference to the original source; and passing off the ideas, arrangement, or conclusions of a published author as your own.

Doing Research.

Doing research for a course paper is like so many other things in life; it's more fun when you're good at it. Completing a research paper can be downright pleasurable once you have mastered the step-by-step process of library research.

Getting Started.

You've undoubtedly heard this lecture before but here it is again: Get an early start. Start to work on your paper during the very first few weeks of class by selecting a topic and investigating possible sources. You will be glad you did.

The earlier you go to the library, the less likely it is that the books you need will have been checked out by someone else, and the more likely it is that you can get the books you need that are not there through interlibrary loan. It's also important to get an early start because school work tends to accumulate toward the end of a semester. There are final exams to take, course projects and other assignments to finish, not to mention papers for other classes to write, which means that if you finish your research paper ahead of time, you'll have more time to do everything else (including rest and recreation).

There is another reason to get an early start. Almost every student experiences a major life event during a four-month-long semester. Illness, heartbreak, financial loss, serious doubts about continuing school, or some other crisis is bound to occur. If your paper is already finished when the storm clouds gather, you will be that much better prepared to keep from getting soaked.

Choosing A Topic.

Sometimes professors assign paper topics, sometimes you're allowed to choose any topic you wish. Most often, you get to choose any topic related to the course that meets with the professor's approval. You can almost always find a topic that interests you by paging through the textbook, browsing in the library, or listening to lecture/class discussion with your paper topic in mind.

You have to live with the topic you choose until the semester is over, so give it serious thought. Don't pick a topic because it sounds interesting or because someone else said it was easy to do. When you have a potential topic, go to the library, look for some books, read an article or two, and request your professor's advice. If your idea still appears promising, chances are you have found a suitable topic.

Remember, library research is easier to do the more specific your topic. Finding, and more importantly, limiting your search for useful sources is less difficult on a narrower topic like "female police officers' response to domestic violence" than it is on broader topics like "law enforcement and domestic violence" or "women in the criminal justice system."

Finding Sources.

Books are one source of information, but they are not the only source nor are they the best source. The problem with relying only on books is the need for timely information. If you're interested in a current issue or the latest developments in an ongoing issue, you are likely to have a problem with books because much of the material is probably out-of-date.

Articles in scholarly journals are one of the best sources of information. Scholarly journal articles represent the most current, most authoritative material on a wide variety of criminal justice issues, and they provide primary research material in brief, manageable amounts. *American Sociological Review, American Journal of Sociology, Journal of Health and Social Behavior, Social Problems,* and *Social Psychology* are just a few of the many sociological journals available. And, depending on your topic, you may be able to find relevant information in the journals of other disciplines, such as political science, psychology, or history (see "How to Find Sources for Research Papers," by Paul Knepper).

Reading to Write.

As you look for sources, discipline yourself. Stick to your topic and minimize the time you spend browsing. Six hours in the library won't get you closer to a completed paper if you spend four of them thumbing through the newspaper, chatting with friends, or snoozing.

Read critically. Before reading an entire book, read the introduction to find the author's credentials. Does the author work at a university? For the government? Also try to evaluate the author's sources. Is the book based on firsthand research, or is it chiefly based on secondary sources?

You should also read with your purpose in mind--writing a paper. So skim the book or article first for material on your topic. If you find valuable information, read further; if not, look for another article. Do not read an entire book or article that doesn't really relate to your topic. By reading selectively, you will be able to cover a dozen times more potential sources than if you read every single word of everything you come across.

Taking Notes.

Take notes on the sources you decide to use. Your notes should include important bits of information, the key arguments of different authors, and memos to yourself about how to arrange material and about specific points you intend to discuss. Unless the material you want can be transcribed quickly, it is usually best to check out the book or xerox the article. (Besides saving time in the library, once you have a photocopy you can highlight the relevant parts.)

Your notes should also include complete bibliographic information for each of your sources;" complete bibliographic information" means everything other people need to know to find it for themselves. For a book, you need to write down the author's name, title of the book, year of publication, publisher, and city of publication. For a journal article, you need to record the author, title of the article, name of the journal, year of publication, volume number, and page numbers of the article.

Needless to say, the better organized and more thorough your notes are, the less time you'll spend writing later on.

Writing the Paper.

After a sufficient period of time, take stock of your library work. Collect your sources, read your notes, and think about your topic. When you have something to say, and a general idea about the way you plan to organize your sources to support your statement, it's time to begin writing.

Although you probably have had a teacher at some point in your educational career who assigned separate grades for form and content, it is actually impossible to separate one from the other. The way any writer organizes her or his ideas on the page reveals how those ideas are organized in her or his mind. A disorganized and confusing paper exposes a writer whose thinking is disorganized and confused, either because she or he is unwilling to do, or is incapable of doing, something better. So pay close attention to both what you say and how you say it.

Every good research paper develops a thesis, displays effective organization, and provides adequate documentation.

Developing a Thesis.

The word thesis sounds scholastic and textbookish. The word sometimes refers to the lengthy paper that a graduate student must write before receiving an advanced degree. Generally, it refers to the most basic part of everything people in college write. Every article, every book, every course paper has a thesis.

A thesis is the author's main point. It is the overall statement, the fundamental argument, the primary reason for writing the paper. The reason the same word is used to describe what a candidate for a graduate degree must produce and what a college student does in a research paper is that both of them basically involve the same thing: taking a stand and defending it.

It's so important to have a thesis in a research paper because a thesis is like the plot in a novel. The plot in a literary work is the main story or plan devised by the author to capture and hold the reader's interest. The plot is what gives the book a provocative beginning, an exciting middle, and a logical end. Similarly, the thesis in a research paper is what makes the paper worth reading. Without a thesis a research paper doesn't make sense.

But writing a college paper is different from writing a novel in one very important respect. The novelist saves the most meaningful part of the story until the very end in order to provide a satisfying conclusion. In a murder mystery, the writer waits to tell the reader whodun-it until the last few pages. A college paper is not the same--don't wait till the conclusion to reveal your thesis. You're not writing a mystery! Your thesis should be clearly stated in the introduction. The reader should know what point you intend to make right from the opening paragraphs and should be able to understand how each of the paragraphs that follows relates to that point.

Organizing Your Thoughts.

How effective you are at getting your message across to the reader depends on how well you organize your paper. If your thesis is not clearly developed, your professor will most likely miss it and assign a grade worthy of a novel without a plot. Even if your thesis is brilliant, no one will know unless you are able to communicate it effectively.

Use subheadings, or "subheads" as they are sometimes called, to identify the major sections of your paper. Subheadings cut the thesis you serve up into bite-sized pieces; they divide your overall presentation into digestible portions. For difficult material, use three levels; for simple material, a single level will do. (I'm using two levels in these guidelines: "Doing Research" and "Writing the Paper" indicate two of the major parts; "Getting Started," "Choosing a Topic," and so on subdivide one of those major parts.)

Whatever you write, use paragraphs. If your paragraphs hang together, your paper will hang together, so it is important to construct them properly. Although paragraphs may be as short as two sentences or as long as a page, they must always express a single, coherent idea. Every paragraph should begin with a topic sentence, which tells the reader what the paragraph is about, and end with a sentence that mentally prepares the reader for the next paragraph.

Use transitions to weld together the sections and paragraphs that fill out your paper. Transitions are words, phrases, sentences, and paragraphs that link your ideas to one another to form a recognizable structure. They may be as brief as "nevertheless" or "however" or as long as a summary paragraph, but they are all vital connections that give unity to composition. Transitions inform the reader where you are, remind the reader where you've been, and signal the reader as to where you plan to go.

Documenting Your Ideas.

In a research paper, everything you say must be documented. Documenting your ideas means supporting them or backing them up with references to sources the reader can investigate for herself or himself. References within college research papers are organized by means of accepted reference formats, such as Chicago Manual style, APA style, and MLA style. (Some of the most widely used style manuals are listed at the end of these guidelines.) Whatever style you choose, consult a style manual, follow it closely, and be consistent.

Often it is a good idea to "tag" your sources as you write. In casual conversation, it's permissible to say "they passed a law" or "they changed that policy," but in academic writing you need to explain who "they" refers to. So, rather than writing "they passed a law," you might write something like: "According to the U.S. Supreme Court's decision in the case of...," or "James Q. Wilson, a political scientist at UCLA, argues that...." Tagging builds credibility because it verifies your facts and opinions by giving information about the source right alongside them.

Editing.

Writing is a process, not an act. A well-written research paper doesn't just unexpectedly fall out of a person's head during the commercials while watching TV one evening. A top-notch research paper demands writing and rewriting; it requires finding better words, altering sentences, and reordering paragraphs. The sooner you get used to the idea that a paper requires more than a single draft, the sooner written communications skills will become a valuable part of your resume.

An excellent research paper shows evidence of editing. Editing is what makes the difference between a feature film and a home movie. Home movies are boring because they amount to an assortment of unrelated scenes all jumbled together. Unless you're one of the stars, it's difficult to know what is going on. Films--the kind you pay money to see in a theatre--have been edited to make sure they tell a comprehensible story. The action sequences, the romance scenes, and the special effects are deliberately and carefully arranged to entertain the viewer. The same is true of a properly edited research paper: The writer has taken the time to make sure that the statistics, court decisions, arguments, facts, and opinions all fit together.

Preparing the Final Copy.

The final appearance of your research paper means nothing and everything. A paper produced with expensive computer software and bound with a transparent plastic report cover won't get you a good grade if the substance of your paper isn't good to begin with. On the other hand, a handwritten copy scribbled with a pen nearly out of ink on paper torn from a spiral-bound notebook will seriously detract from even the finest academic scholarship.

The standard research paper written by a professional is typed (or if a word processor is used, printed) on letter-size paper. The text is double-spaced, and the pages are simply stapled together. Professional papers--research articles by professors, reports by government officials, even the opinions of Supreme Court justices--are all made to look this same way prior to publication.

So, if you want to produce a professional-looking paper, do it the same way. Type or print your paper on letter-size paper. (Handwritten work is never acceptable for a research paper in a college course.) Double-space the text throughout, and clearly label major sections, appendices, and tables. Be sure to include a title page and attach your reference list. Do not use legal-size paper or onionskin (erasable bond) paper. Avoid plastic report covers--they make your paper hard to read, tend to fall apart, and give you away as a rank amateur.

One final note. If you hire a typist, you must still proofread your paper. You, the person who wrote the paper (not the person who typed the paper), are responsible for the finished product.

Style Manuals

The Chicago Manual of Style. 13th ed. 1982. Chicago: University of Chicago Press.

MLA *Handbook for Writers of Research Papers*. 3rd ed. 1988. New York: Modern Language Association of America.

Publication Manual of the American Psychological Association. 3rd ed. 1983. Washington, DC: American Psychological Association.

Webster's Standard Style Manual. 1985. Springfield, MA: Merriam-Webster.

HOW TO WRITE A BOOK REVIEW

A book review is not the same as a book report. You may have written a book report for a class in high school or for a college English course. Your main purpose for such an assignment was to show your instructor that you actually read the book, and maybe you added some comments about parts of the book you found interesting. A book review has a distinctly different purpose. Your job in writing a review is to describe the major themes in the book and discuss the issues they present.

Writing a good book review requires careful reading and thoughtful writing. It is important to read the book with your purpose in mind--writing a review. So take notes as you read in addition to marking important portions. Your notes will likely include a clear statement of the author's purpose, some ideas about the structure and techniques used to achieve that purpose, and perhaps some references to books by other authors who discuss similar issues.

In a book review, your task is to critique the author's work. The word critique does not mean to criticize, find fault with, or tear down the author. Nor does it mean judging the book according to how you feel about the topic, or presenting what you know about the author's topic. Critique means "critical assessment," and it's the most important aspect of a book review.

Specifically, a book review that offers a critique answers three questions: What is the author's purpose for writing the book? How well does the book achieve that purpose? How significant is the total work considering other information on the same topic? These three questions suggest three major sections that are part of every book review: summary, analysis, and evaluation.

Summary

Every book review begins with an accurate, well-organized summary of the author's main ideas. Sumarize means to paraphrase or restate briefly in your own words. Your restatement should be framed around the author's thesis or main idea; it should restate the author's purpose for writing the book. Identify the primary question the author is concerned with, the central points the author makes, and the gist of the information the author presents.

You may capture the tone of the book, perhaps with a well-chosen quote, but you should not present any of your own ideas, thoughts, or opinions. An effective summary is marked by succinctness, clarity, and objectivity.

Analysis

The middle portion of a book review contains analysis. To analyze is to assess how well the author accomplishes the purpose for the book. Examine the method or technique used to achieve the author's central argument. What makes the arguments rational and convincing? How does the evidence presented support the author's conclusions?

Offer any suggestions or changes you may be able to provide. If you use quotes in this portion of your review, be sure to make them brief, and use them only to illustrate a specific point you are making.

Evaluation

The final portion of a book review is evaluation. To evaluate the book means to assess the significance of the total work. Now you offer an informed opinion about the book as a whole. In order to support your evaluation, refer to comments expressed earlier in the analysis section of your review. Be sure that your remarks are qualified, learned, and thoughtful.

Never make such statements as "This is a dumb book. It was boring," or "I just hated this book, I don't know why." Your specific objective here is to evaluate the contribution the book makes within the wider body of information about the topic and present the basis for your conclusion.

When you are satisfied that you have sufficiently summarized, analyzed, and evaluated the book, edit your review. To edit means to bring about conformity by rearranging, altering, and assembling. Read your review from beginning to end to make sure that your writing is coherent. You may need to delete misleading words, add transition sentences, or reorder your paragraphs. Make sure that your sentences are clear, that your paragraphs follow a logical direction, and that your words adequately express your thoughts and ideas.

After you have edited your review--and before you give it to your instructor for credit --proofread it. Correct any spelling, grammar, or punctuation mistakes that remain.

APPENDIX B

Maps

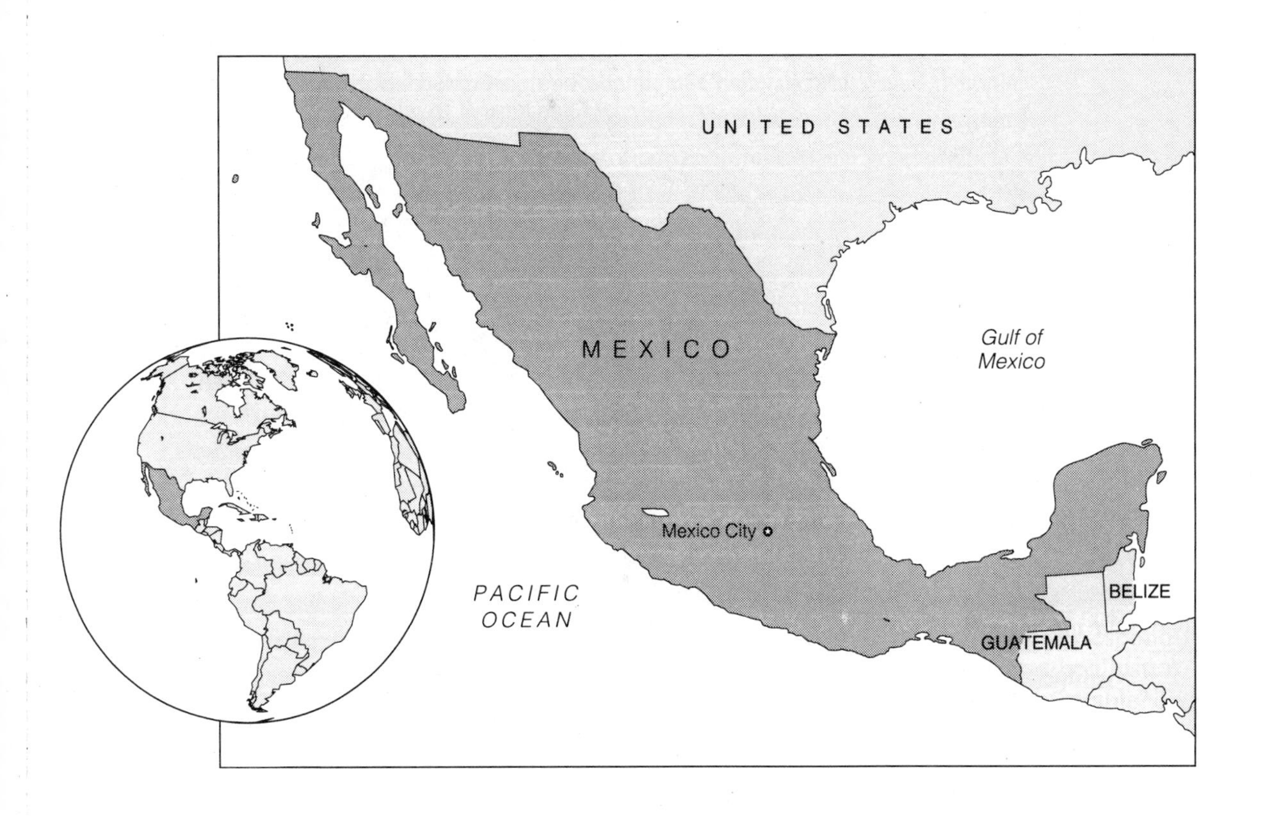
UNITED STATES
MEXICO
Gulf of
Mexico
Mexico City
PACIFIC
OCEAN
BELIZE
GUATEMALA

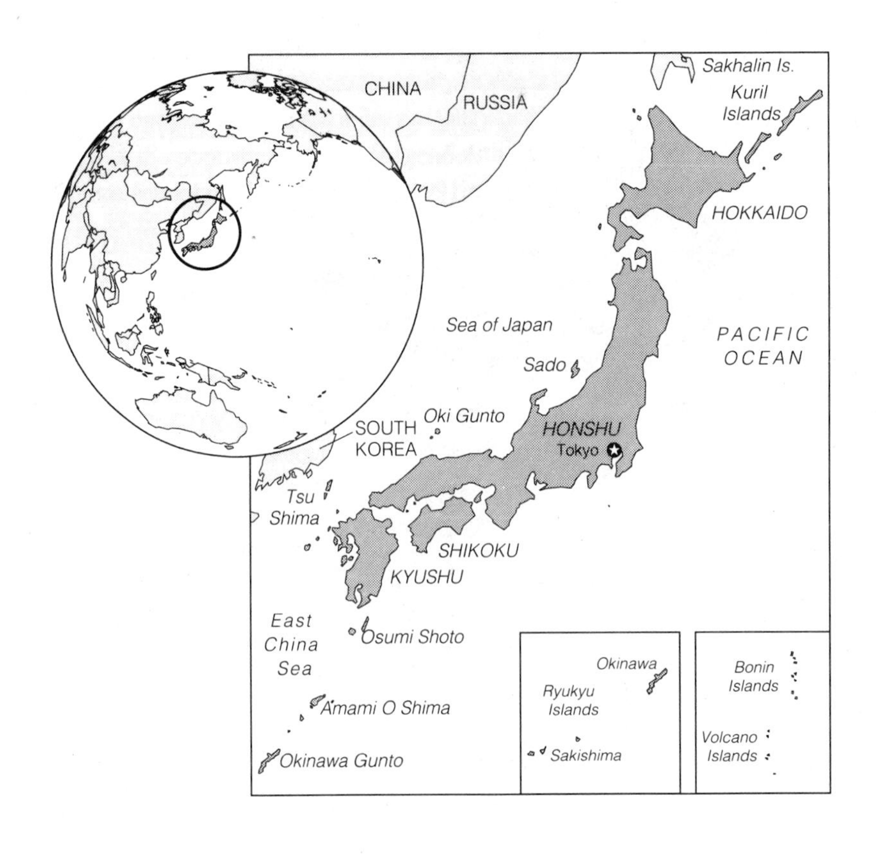
CHINA
RUSSIA
Sakhalin Is.
Kuril Islands
HOKKAIDO
Sea of Japan
PACIFIC OCEAN
Sado
Oki Gunto
SOUTH KOREA
HONSHU
Tokyo
Tsu Shima
SHIKOKU
KYUSHU
East China Sea
Osumi Shoto
Amami O Shima
Okinawa Gunto
Okinawa
Ryukyu Islands
Sakishima
Bonin Islands
Volcano Islands

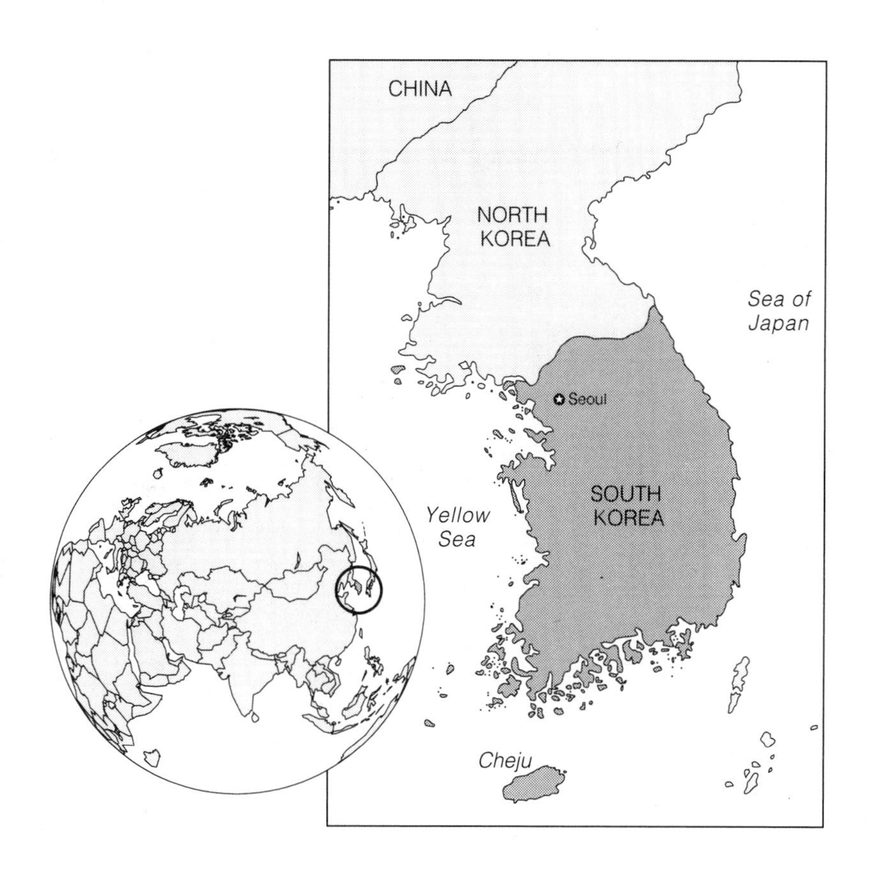
CHINA
NORTH KOREA
Sea of Japan
Seoul
Yellow Sea
SOUTH KOREA
Cheju

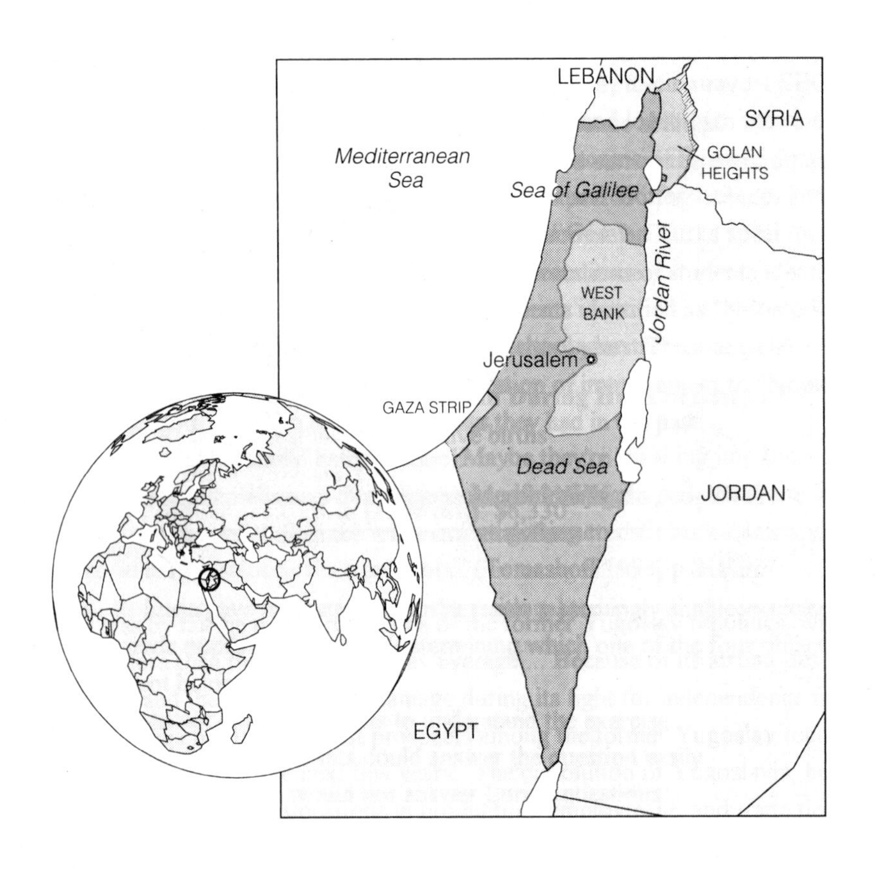
LEBANON
SYRIA
Mediterranean Sea
GOLAN HEIGHTS
Sea of Galilee
Jordan River
WEST BANK
Jerusalem
GAZA STRIP
Dead Sea
JORDAN
EGYPT

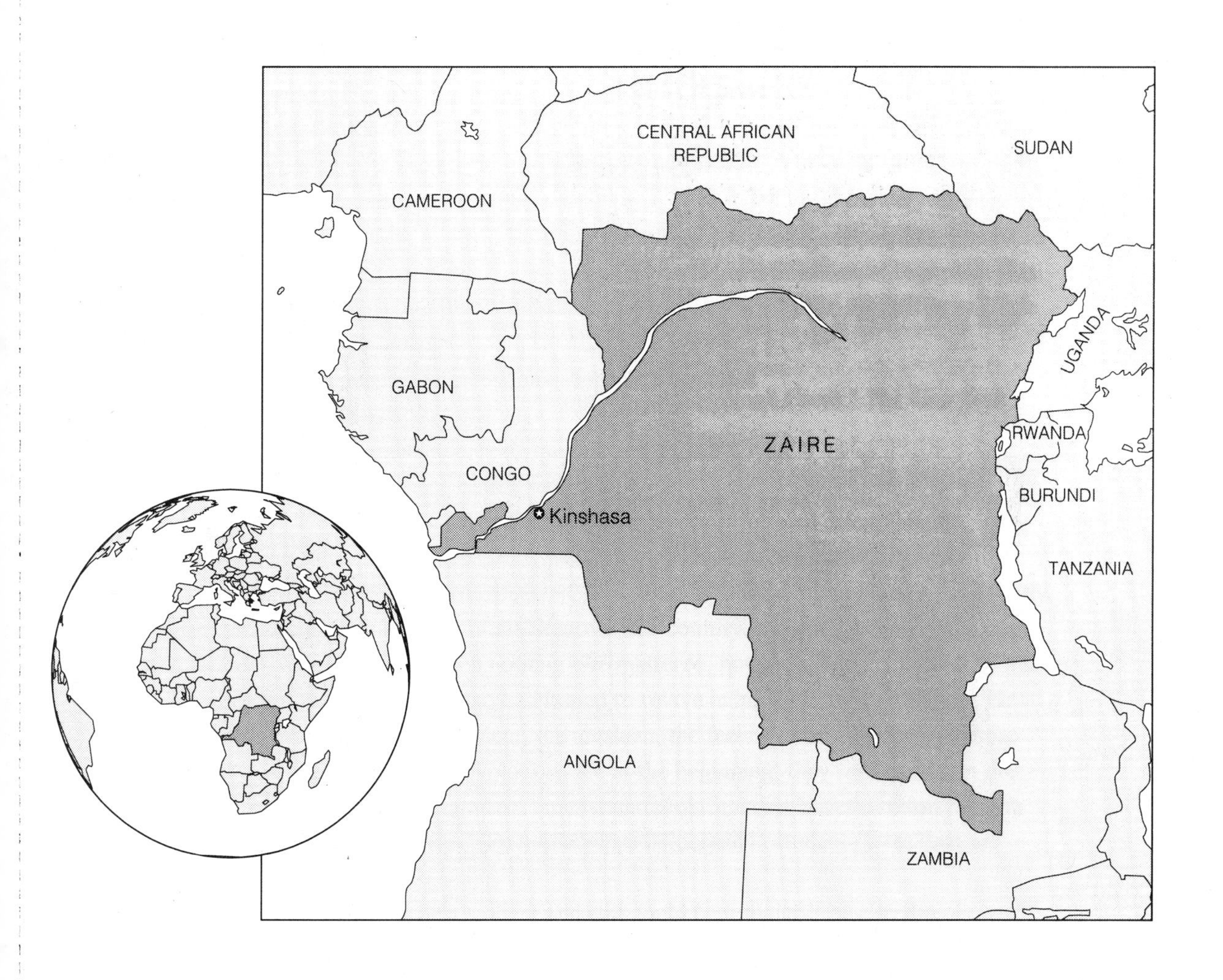
CENTRAL AFRICAN
REPUBLIC
SUDAN
CAMEROON
GABON
CONGO
Kinshasa
ZAIRE
UGANDA
RWANDA
BURUNDI
TANZANIA
ANGOLA
ZAMBIA

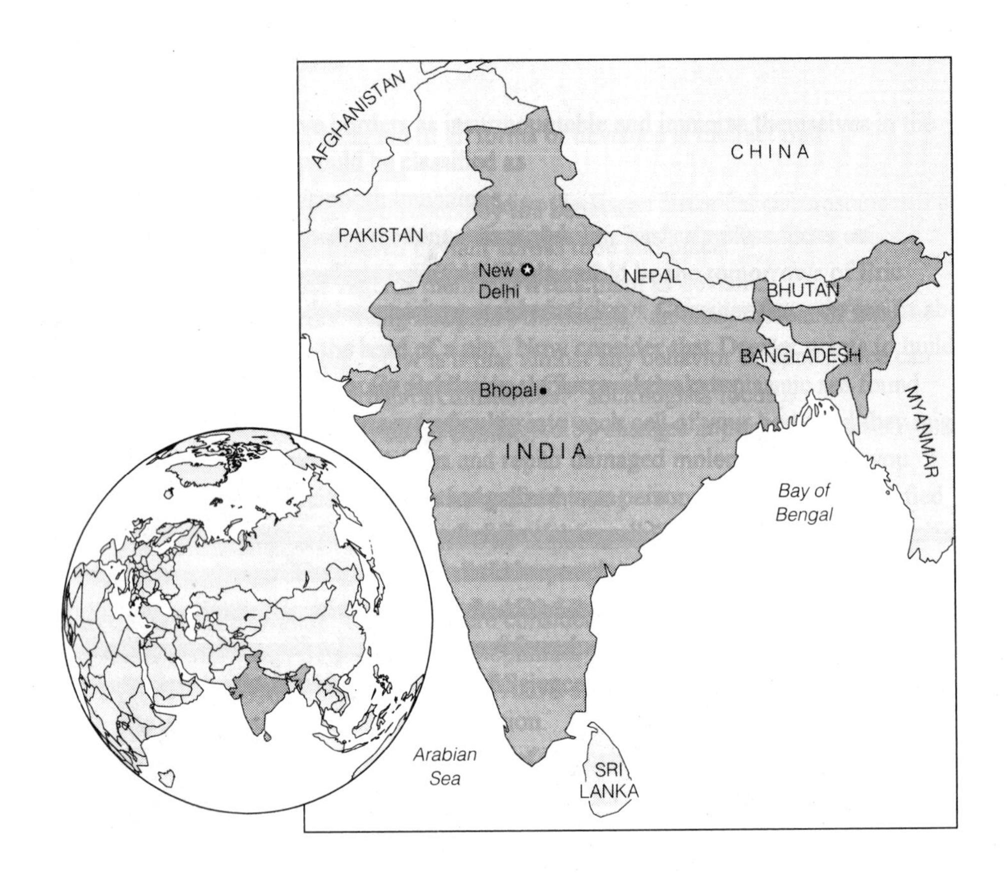
AFGHANISTAN
CHINA
PAKISTAN
New Delhi
NEPAL
BHUTAN
BANGLADESH
Bhopal
INDIA
MYANMAR
Bay of Bengal
Arabian Sea
SRI LANKA

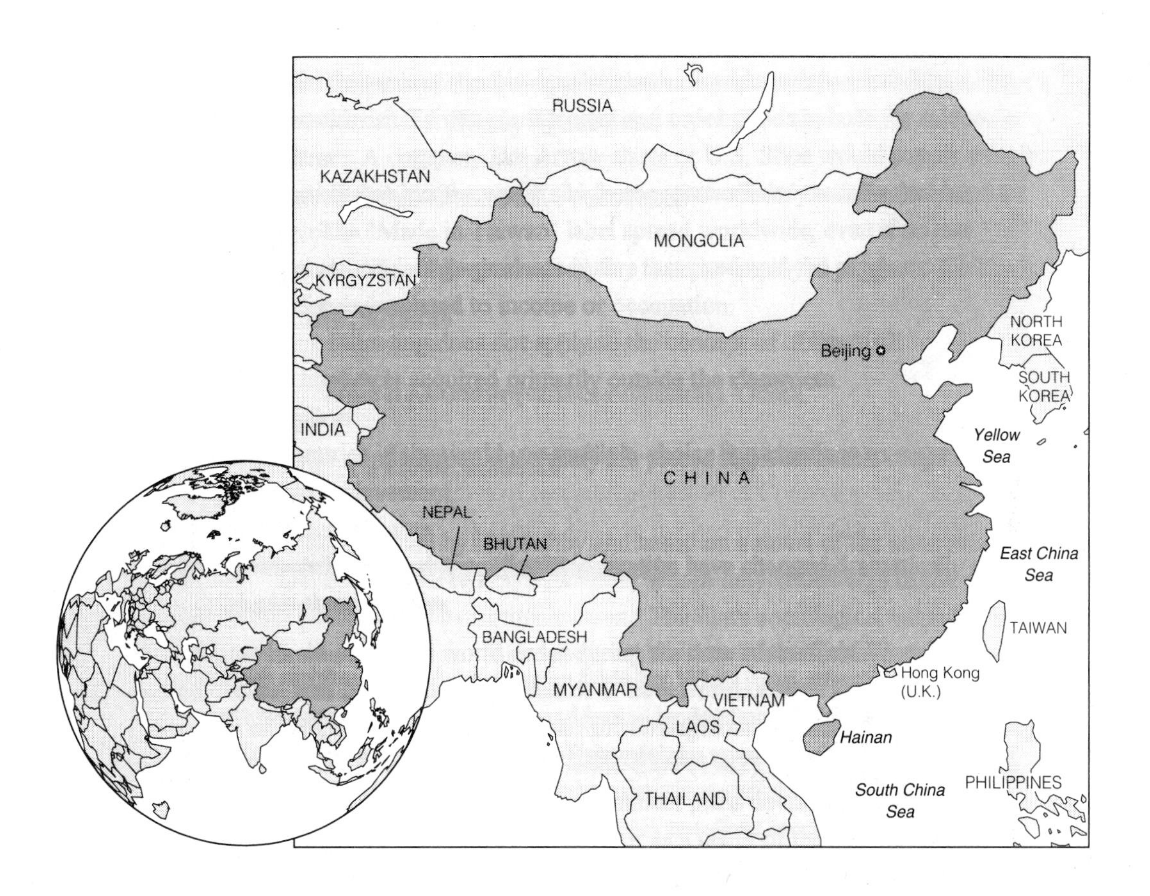
RUSSIA
KAZAKHSTAN
MONGOLIA
KYRGYZSTAN
NORTH KOREA
Beijing
SOUTH KOREA
INDIA
Yellow Sea
CHINA
NEPAL
BHUTAN
East China Sea
TAIWAN
BANGLADESH
Hong Kong (U.K.)
MYANMAR
VIETNAM
LAOS
Hainan
South China Sea
PHILIPPINES
THAILAND

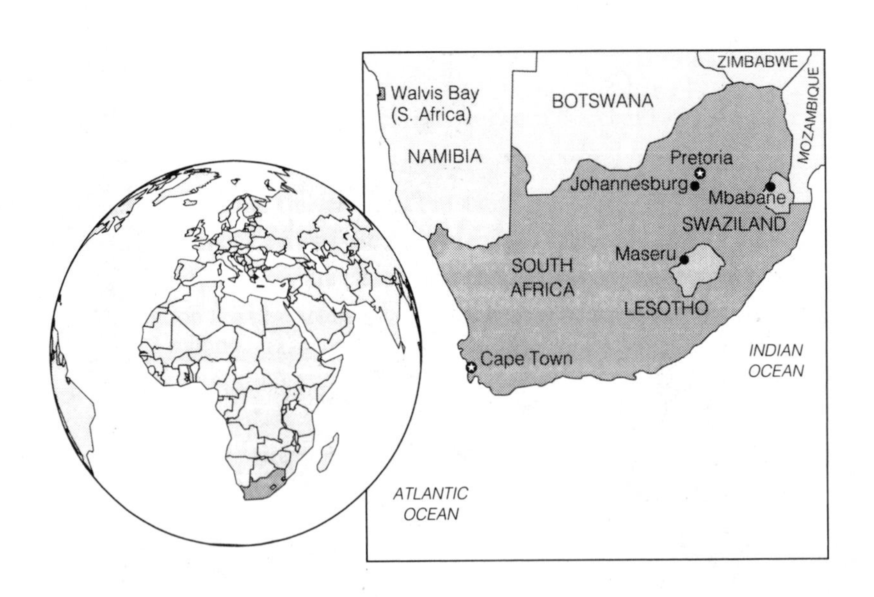
ZIMBABWE
Walvis Bay
(S. Africa)
BOTSWANA
MOZAMBIQUE
NAMIBIA
Pretoria
Johannesburg
Mbabane
SWAZILAND
Maseru
SOUTH
AFRICA
LESOTHO
Cape Town
INDIAN
OCEAN
ATLANTIC
OCEAN

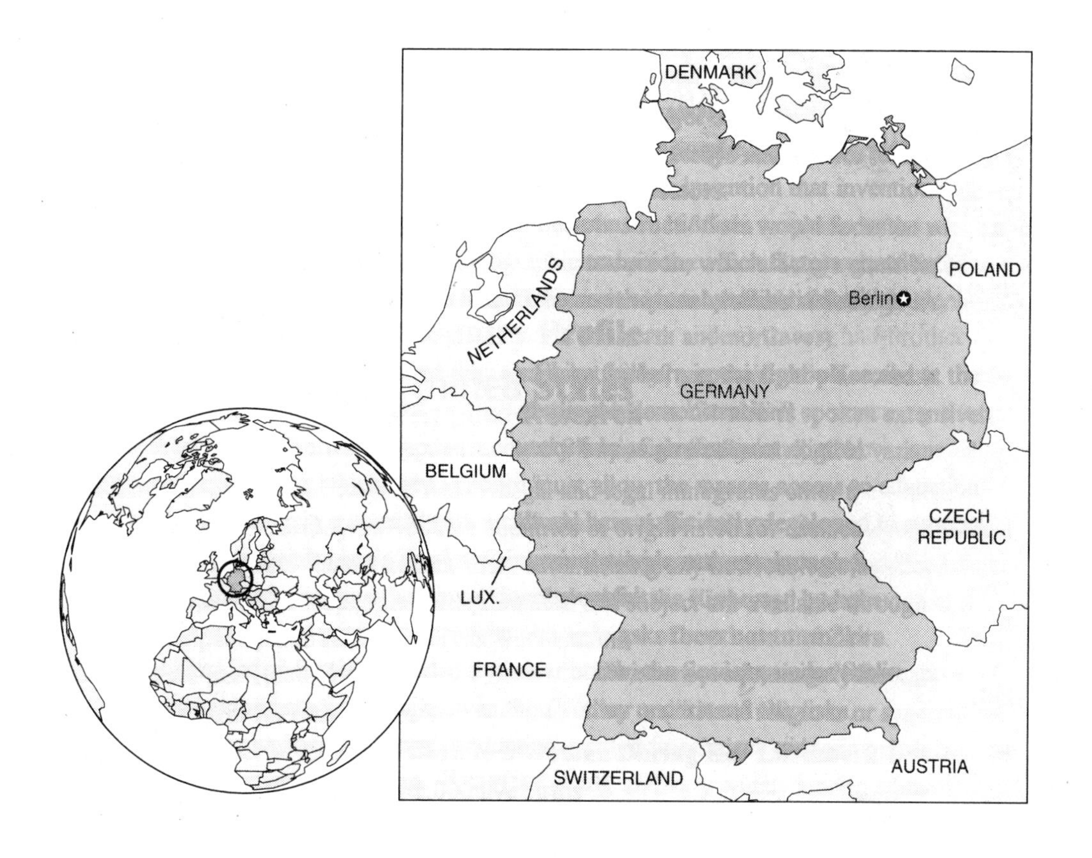
DENMARK
POLAND
Berlin
NETHERLANDS
GERMANY
BELGIUM
CZECH
REPUBLIC
LUX.
FRANCE
SWITZERLAND
AUSTRIA

AUSTRIA
HUNGARY
ROMANIA
Ljubljana
SLOVENIA
Zagreb
CROATIA
VOJVODINA
Novi Sad
BOSNIA
AND
HERZEGOVINA
Belgrade
ITALY
Dubrovnik
Sarajevo
SERBIA
Adriatic Sea
MONTE-
NEGRO
Titograd
Pristina
KOSOVO
BULGARIA
Skopje
MACEDONIA
ALBANIA
GREECE
Mediterranean
Sea

VENEZUELA
GUYANA
FRENCH
GUIANA
COLOMBIA
SURINAME
BRAZIL
PERU
Brasília
BOLIVIA
PARAGUAY
ATLANTIC
OCEAN
ARGENTINA
URUGUAY

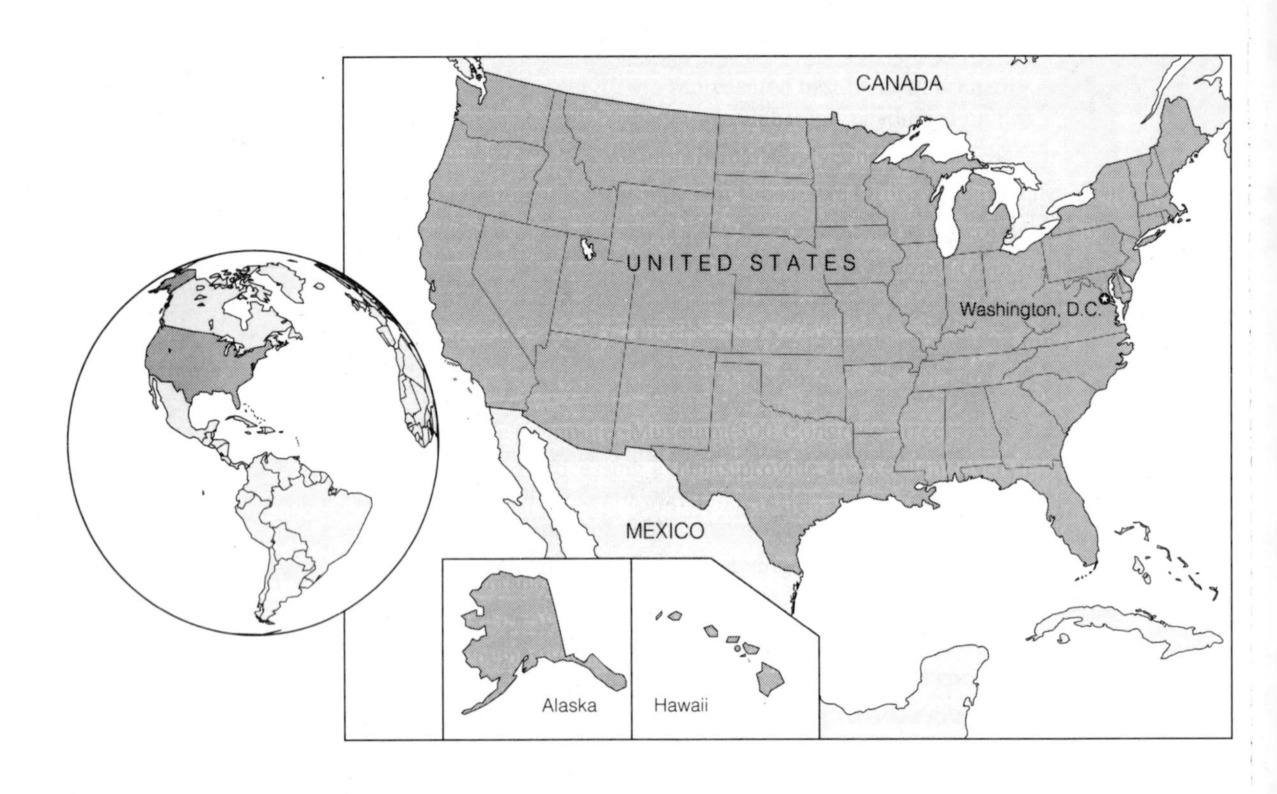
CANADA
UNITED STATES
Washington, D.C.
MEXICO
Alaska
Hawaii

Mediterranean
Sea
LEBANON
Beirut
SYRIA
GOLAN
HEIGHTS
ISRAEL

ATLANTIC OCEAN
ARCTIC OCEAN
LUX.
SWITZ.
WEST GERMANY
EAST GER.
R.S.F.S.R.
FINLAND
ITALY
AUST.
CZECH
POLAND
ESTONIAN S.S.R.
LATVIAN S.S.R.
LITHUANIAN S.S.R.
HUNG.
YUGOSLAVIA
BYELORUSSIAN S.S.R.
UNION OF SOVIET SOCIALIST REPUBLICS
ALBANIA
ROMANIA
UKRAINIAN S.S.R.
GREECE
BULGARIA
MOLDAVIAN S.S.R.
RUSSIAN SOVIET FEDERATED SOCIALIST REPUBLIC
GEORGIAN S.S.R.
TURKEY
ARMENIAN S.S.R.
KAZAKH S.S.R.
AZERBAIJAN S.S.R.
NORTH KOREA
MONGOLIA
TURKMEN S.S.R.
UZBEK S.S.R.
KIRGIZ S.S.R.
IRAN
TADZHIK S.S.R.
SOUTH KOREA
AFGHANISTAN
CHINA
PACIFIC OCEAN
PAKISTAN